Praise for *The Biology of Belief*

"I read *The Biology of Belief* when it first came out. It was a pioneering book and gave a much needed scientific framework for the mind/body/spirit connection. Bruce's insights and research created the basis of the epigenetic revolution that is now laying the foundation for a consciousness-based understanding of biology. We are all indebted to him."

— Deepak Chopra, M.D., F.A.C.P.,
co-author with Rudolph Tanzi of *Super Genes: Unlock the Astonishing Power of Your DNA for Optimum Health and Well-Being*

"Bruce Lipton's book is the definitive summary of the new biology and all it implies. It is magnificent, profound beyond words, and a delight to read. It synthesizes an encyclopedia of critical new information into a brilliant yet simple package. These pages contain a genuine revolution in thought and understanding, one so radical that it can change the world."

— Joseph Chilton Pearce, Ph.D.,
author of *Magical Child* and *Evolution's End*

"Bruce Lipton's delightfully written *The Biology of Belief* is a much needed antidote to the 'bottom-up' materialism of today's society. The idea that DNA encodes all of life's development is being successfully employed in genetic engineering. At the same time, the shortfall of this approach is becoming evident. *The Biology of Belief* is a review of a quarter-century of pioneering results in Epigenetics, heralded by *The Wall Street Science Journal* in mid-2004 as an important new field. Its personal style makes it eminently readable and enjoyable."

— Karl H. Pribram, M.D., Ph.D.,
(Hon. Multi), professor emeritus, Stanford University

"Dr. Lipton is a genius. His breakthrough discoveries give us tools for regaining the sovereignty over our lives. I recommend this book to anyone who is ready and willing to take full responsibility for themselves and the destiny of our planet."

— LeVar Burton, actor and director

"Bruce Lipton offers new insights and understanding into the interface between biological organisms, the environment—and the influence of thought, perception, and subconscious awareness—on the expression of one's body healing potential. Well-referenced explanations and examples make this book a refreshing 'must read' for the student of the biological, social, and health care sciences. Yet the clarity of the author's presentation makes it an enjoyable read for a general audience."

— Carl Cleveland III, D.C.,
President, Cleveland Chiropractic College

"Dr. Lipton's revolutionary research has uncovered the missing connections between biology, psychology, and spirituality. If you want to understand the deepest mysteries of life, this is one of the most important books you will ever read."

—Dennis Perman, D.C., co-founder, The Master's Circle

"In this paradigm-busting book, Bruce Lipton delivers a TKO to Old Biology. With a left to Darwinian dogma and a right to allopathic medicine, he breaks out of the physicalist box into enlightenment on the mind/body (belief/biology) system. Must read, much fun."

— Ralph Abraham, Ph.D., professor of mathematics,
University of California; author of *Chaos, Gaia, Eros*

"Powerful! Elegant! Simple! In a style that is as accessible as it is meaningful, Dr. Bruce Lipton offers nothing less than the long sought–after 'missing link' between life and consciousness. In doing so, he answers the oldest questions and solves the deepest mysteries of our past. I have no doubt that *The Biology of Belief* will become a cornerstone for the science of the new millennium."

— Gregg Braden,
best-selling author of *The God Code* and *The Divine Matrix*

"I finished reading this book with the same sense of profound respect I have when I am with Bruce Lipton—that I have been touched by a revolutionary sense of the truth. He is both a scientist and a philosopher; a scientist in that he provides us with tools to alter cultural consciousness and a philosopher because he challenges our beliefs about the very nature of our perceived reality. He is helping us create our own futures."

— Guy F. Riekeman, D.C.,
President, Life University and College of Chiropractic

"*The Biology of Belief* is a milestone for evolving humanity. Dr. Bruce Lipton has provided, through his amazing research and in this inspiring book, a new, more awakened science of human growth and transformation. Instead of being limited by the genetic or biological constraints that humanity has been programmed to live by, humanity now has before it a way of unleashing its true spiritual potential with the help of simply transformed beliefs guided by 'the gentle and loving hand of God.' A definite must read for those dedicated to the mind/body movement and to the true essence of healing."

— Dr. John F. Demartini, best-selling author of
Count Your Blessings and *The Breakthrough Experience*

"In a world of chaos, Dr. Lipton brings clarity to mankind. His work is thought-provoking, insightful, and will hopefully lead people to ask better quality questions in their lives and to make better decisions. One of the most exciting books I have read, this is a must read."

— Brian Kelly, D.C., President, New Zealand College of
Chiropractic; President, Australian Spinal Research Foundation

"Finally, a compelling and easy-to-understand explanation of how your emotions regulate your genetic expression! You need to read this book to truly appreciate that you are not a victim of your genes but instead have unlimited capacity to live a life overflowing with peace, happiness, and love."

— Joseph Mercola, D.O., Founder of
www.mercola.com, world's most visited natural-health site

"This book is an absolute must read if you want to know, from a scientific viewpoint, that your lifestyle is in control of your health rather than your genetics. From a scientific viewpoint, Lipton demonstrates that the mind is more powerful than drugs to regain our health. The information reveals that your health is more your responsibility than just being a victim of your genes. When I started reading this book, I could not stop until it was finished."

— M. T. Morter, Jr., D.C.,
founder, Morter Health System;
developer of the B.E.S.T. Technique

"This is a courageous and visionary book that provides solid evidence from quantum biology to dispel the myth of genetic determinism—and implicitly, victimhood. Dr. Bruce Lipton brings a solid scientific mind to not only inform but to transform and empower the reader with the realization that our beliefs create every aspect of our personal reality. A provocative and inspiring read!"

— Lee Pulos, Ph.D., A.B.P.P.,
professor emeritus, University of British Columbia;
author of *Miracles and Other Realities* and *Beyond Hypnosis*

"History will record *The Biology of Belief* as one of the most important writings of our time. Bruce Lipton has delivered the missing link between the understandings of biomedicine of the past and the essentials of energetic healing of the future. His complex insights are expressed in a readily understandable fashion with a style that welcomes the scientist and the nonscientist on an equal footing. For anyone interested in health, the well-being of the species, and the future of human life, *The Biology of Belief* is a must read. The implications of the perspectives outlined have the potential to change the world as we know it. Bruce Lipton's understandings—and his concise expression of them—are sheer genius."

— Gerard W. Clum, D.C.,
President, Life Chiropractic College West

THE BIOLOGY OF BELIEF

THE BIOLOGY OF BELIEF

Unleashing the Power of Consciousness,
Matter & Miracles

Bruce H. Lipton, Ph.D.

HAY HOUSE, INC.
Carlsbad, California • New York City
London • Sydney • New Delhi

Published in the United States by: Hay House, Inc.: www.hayhouse.com®
Published in Australia by: Hay House Australia Pty. Ltd.: www.hayhouse.com.au
Published in the United Kingdom by: Hay House UK, Ltd.: www.hayhouse.co.uk
Published in India by: Hay House Publishers India: www.hayhouse.co.in

Indexer: Susan Edwards
Cover design: ©2004 Robert Mueller
Interior design: Nick C. Welch

Library of Congress Cataloging-in-Publication Data for the original edition

Lipton, Bruce H.
  The biology of belief : unleashing the power of consciousness, matter & miracles
/ Bruce H. Lipton, Ph.D. -- 10th anniversary edition.
    pages cm
  Includes bibliographical references and index.
  ISBN 978-1-4019-4891-7 (hardcover : alk. paper) 1. Cytology. 2. Molecular biology. 3. Genetic psychology. 4. Psychophysics. I. Title.
  QH581.2.L56 2015
  599.93'5--dc23
                                    2015010850

Tradepaper ISBN: 978-1-4019-5247-1

25 24 23 22 21 20 19
1st edition, October 2015
2nd Edition, October 2016

Printed in the United States of America

SUSTAINABLE FORESTRY INITIATIVE

Certified Chain of Custody
Promoting Sustainable Forestry
www.sfiprogram.org
SFI-01268

SFI label applies to the text stock

This book is dedicated to . . .

GAIA

The Mother of Us All
May She forgive us our trespasses.

To my own mother, Gladys,
who has continually encouraged and supported me
while being patient for the twenty years
it took to get this book out.

To my daughters, Tanya and Jennifer,
beautiful women of the world who have always been there
for me . . . no matter how weird things had become.

And especially to my darling, Margaret Horton,
my best friend, my life partner, my love.
May we continue on our joyous quest
to live happily ever after!

Contents

Prologue

"If you could be *anybody*, who would you be?" I used to spend an inordinate amount of time pondering that question. I was obsessed with the fantasy of changing my identity because I wanted to be anybody *but* me. I had a good career as a cell biologist and medical school professor, but that didn't make up for the fact that my personal life was, at best, a shambles. The harder I tried to find happiness and satisfaction in my personal life, the more dissatisfied and unhappy I became. In my reflective moments, I resolved to surrender to my unhappy life. I decided that fate had dealt me a bad hand, and I should simply accept it. Que sera, sera.

In the fall of 1985, my depressed, fatalistic attitude changed in one transformational moment. I had resigned my tenured position at the University of Wisconsin's School of Medicine and was teaching at an offshore medical college in the Caribbean. Because the school was so far from the academic mainstream, I had the opportunity to think outside the rigid parameters of *belief* that prevail in conventional academia. Far from the ivory towers, isolated on an emerald island in the deep azure Caribbean Sea, I experienced a scientific epiphany that shattered my *beliefs* about the nature of life.

My life-changing moment occurred while I was reviewing my research on the mechanisms by which cells control their physiology and behavior. Suddenly I realized that a cell's life is fundamentally controlled by the physical and energetic environment with only a small contribution by its genes. Genes are simply molecular blueprints used in the construction of cells, tissues, and organs. The environment serves as a "contractor" who reads and engages those genetic blueprints and is ultimately responsible for the character of a cell's life. It is a single cell's "awareness" of the environment that primarily sets into motion the mechanisms of life.

As a cell biologist I knew that my insights had powerful ramifications for my life and the lives of all human beings. I was acutely aware that each of us is made up of approximately 50 trillion single cells. I had devoted my professional life to better understanding single cells because I knew then and know now that the better we understand single cells the better we can understand the community of cells that comprises each human body and that if single cells are controlled by their awareness of the environment so too are we trillion-celled human beings. Just like a single cell, the character of our lives is determined not by our genes but by our responses to the environmental signals that propel life.

On the one hand, this new understanding of the nature of life was a jolt. For close to two decades I had been programming biology's Central Dogma—the *belief* that life is controlled by genes—into the minds of medical students. On the other hand, my new understanding was not a complete surprise. I had always had niggling doubts about genetic determinism. Some of those doubts stemmed from my eighteen years of government-funded research on cloning stem cells. Though it took a sojourn outside of traditional academia for me to fully realize it, my research at that time (1985) offered incontrovertible proof that biology's most cherished tenets regarding genetic determinism were fundamentally flawed.

My new understanding of the nature of life not only corroborated my stem cell research but also, I realized, contradicted another *belief* of mainstream science of that time. I had been propounding to my students—the *belief* that allopathic medicine is the only kind of medicine that merits consideration in medical school. By finally giving the energy-based environment its due, it provided for a grand convergence uniting the science and practice of allopathic medicine, complementary medicine, and the spiritual wisdom of ancient and modern faiths.

On a personal level, I knew at the moment of insight that I had gotten myself stuck simply by *believing* that I was fated to have a spectacularly unsuccessful personal life. There is no doubt that human beings have a great capacity for sticking to false *beliefs* with great passion and tenacity, and hyper-rational scientists are not immune. Our well-developed nervous system, headed by our big brain, is testament that our awareness is far more complicated than that of a single cell. When our uniquely human minds get involved, we can choose to perceive the environment in different ways, unlike a single cell whose awareness is more reflexive.

I was exhilarated by the new realization that I could change the character of my life by changing my *beliefs*. I was instantly energized because I realized that there was a science-based path that would take me from my job as a perennial "victim" to my new job as "co-creator" of my destiny.

It has been thirty years since that magical night in the Caribbean when I had my life-changing moment of insight and ten years since I published the first edition of *The Biology of Belief*. In the intervening years, and particularly in the last decade, biological research has corroborated the knowledge I gained on that early morning in the Caribbean. We are living in exciting times, for science is in the process of shattering old myths and rewriting a fundamental *belief* of human civilization. The *belief* that we are frail, biochemical machines controlled by genes is giving way to an understanding that we are powerful creators of our lives and the world in which we live.

The times are indeed changin', which is why I'm particularly excited about this tenth-anniversary edition of *The Biology of Belief*. In fact, I thought about a new title for this edition: *The Biology of Belief and Hope*. However, I reconsidered because I like the alliteration of the original title! Nevertheless, during this time of change (despite, I can't deny, a slew of negative news headlines), I am filled with hope.

Hope because the size and enthusiasm of the audiences for my lectures about *The Biology of Belief*, which has been published in thirty-five countries, have grown exponentially.

Hope because more and more professionals, who agree that biomedicine needs to change its drug-focused ways, are coming to my lectures and engaging me in spirited debate.

Hope because I've met so many people who "get" that *The Biology of Belief* isn't just about *individual* empowerment, and it certainly isn't just about me. I was deeply honored to receive the Goi Peace Award in 2009, and I was also thrilled that the President of the Goi Peace Foundation, Hiroo Saionji, made it so clear that though I was the recipient, the award was actually for the "new science" outlined in *The Biology of Belief*: "[This] research . . . has contributed to a greater understanding of life and the true nature of humanity, empowering wide layers of the public to take control of their own lives and become responsible co-creators of a harmonious planetary future."

It is also my sincerest hope that everyone who reads *The Biology of Belief* recognizes that many of the *beliefs* that propel their lives are false and self-limiting. You can take control of your life and set out on the road to health and happiness, and you can band together with others you meet on that road so that humanity can evolve to a new level of understanding and peace.

As for me, I am ever thankful for that moment of insight in the Caribbean, which enabled me to create my now wondrous life. In the last decade, I've traveled around the world several times teaching the New Biology, written two more books—*Spontaneous Evolution* (2009) and *The Honeymoon Effect* (2013)—become a grandfather three times over, and, oh, become a septuagenarian. Instead of slowing down with age, I feel more and more energized by the life I've created, the connections I've made with those who are also dedicated to creating a harmonious planet, and the continuing honeymoon I'm enjoying with Margaret Horton, my best friend, my life partner, my love, as I described her in the first edition's dedication and still describe her now. In short, my life is so much richer and more satisfying that I no longer ask myself: *If I could be anybody, who would I be?* For me, the answer is a no-brainer. I want to be *me*!

Introduction

The Magic of Cells

I was seven years old when I stepped up onto a small box in Mrs. Novak's second-grade classroom, high enough to plop my eye right onto the lens and eyepiece of a microscope. Alas, I was too close to see anything but a blob of light. Finally I calmed down enough to listen to instructions to back off from the eyepiece. And then it happened, an event so dramatic that it would set the course for the rest of my life. A paramecium swam into the field. I was mesmerized. The raucous din of the other kids faded, as did the back-to-school smells of freshly sharpened pencils, new waxy crayons, and plastic Roy Rogers pencil cases. My whole being was transfixed by the alien world of this cell that, for me, was more exciting than today's computer-animated special-effects movies.

In the innocence of my child mind, I saw this organism not as a cell but as a microscopic person, a thinking, sentient being. Rather than aimlessly moving around, this microscopic, single-celled organism appeared to me to be on a mission, though what kind of mission I didn't know. I quietly watched over the paramecium's "shoulder" as it busily comported itself in and around the algal mat. While I was focusing on the paramecium, a large pseudopod of a gangly amoeba began to ooze into the field.

Just then my visit to this Lilliputian world ended abruptly when Glenn, the class bully, yanked me off the step and demanded his turn at the microscope. I tried to get Mrs. Novak's attention, hoping that Glenn's personal foul would get me another minute at the microscope free-throw line. But it was just minutes before lunch time and the other kids in line were clamoring for their turn. Immediately after school, I ran home and excitedly relayed my microscopic adventure to my mother. Using my best second-grade powers of persuasion, I asked, then begged, then cajoled my mother into getting me a microscope, where I would spend hours mesmerized by this alien world that I could access via the miracle of optics.

Later, in graduate school, I advanced to an electron microscope. The advantage of an electron microscope over a conventional light microscope is that it is a thousand times more powerful. The difference between the two microscopes is analogous to the difference between the 25¢ observation telescopes used by tourists to observe scenic vistas and the orbiting Hubble telescope that transmits images of deep space. Entering the electron microscopy suite of a laboratory is a rite of passage for aspiring biologists. You enter through a black revolving door, similar to the ones separating photographic darkrooms from illuminated work areas.

I remember the first time I stepped into the revolving door and began to turn it. I was in darkness between two worlds, my life as a student and my future life as a research scientist. When the door completed its rotation, I was deposited into a large, dark chamber, dimly lit by several red photographic safelights. As my eyes adapted to the available light, I gradually became awed by what stood before me. The red lights were reflecting eerily off the mirrored surface of a massive, foot-thick chromium steel column of electromagnetic lenses that rose to the ceiling in the center of the room. Spreading out on either side at the base of the column was a large control console. The console resembled the instrument panels of a Boeing 747, filled with switches, illuminated gauges, and multicolored indicator lamps. Large tentacle-like arrays of thick power cords, water hoses, and vacuum lines radiated from the base of the microscope like tap roots at the base of an old oak tree. The sound of clanking vacuum pumps and the whir of refrigerated water recirculators filled the air. For all I knew, I had just emerged onto the command deck of the *U.S.S. Enterprise.* Apparently, it was Captain Kirk's day off, for sitting at the console was one of my professors, who was engaged in the elaborate procedure of introducing a tissue specimen into a high-vacuum chamber in the middle of the steel column.

While the minutes passed, I experienced a feeling reminiscent of that day in second grade when I first saw a cell. Finally, a green fluorescent image appeared on the phosphor screen. The presence of darkly stained cells could barely be discerned in the plastic sections, which were enlarged to about thirty times their original size. Then the magnification was increased, one step at a time. First 100X, then 1000X, and then 10,000X. When we finally hit warp drive, the cells were magnified to over 100,000 times their original size. It was indeed *Star Trek,* but rather than entering outer space, we were going deep into inner space where "no man has gone before." One moment I was observing a miniature cell, and seconds later I was flying deep into its molecular architecture.

My awe at being at the edge of this scientific frontier was palpable. So was my excitement when I was made honorary co-pilot. I put my hands on the controls so that I could "fly" over this alien cellular landscape. My professor was my tour guide, pointing out notable cellular landmarks: "Here's a mitochondrion, there's the Golgi body, over there is a nuclear pore, this is a collagen molecule, that's a ribosome."

Most of the rush I experienced came from my vision of myself as a pioneer, traversing territory that had never been seen by human eyes. While the light microscope gave me an awareness of cells as sentient creatures, it was the electron microscope that brought me face to face with the molecules that were the very foundation of life itself. I knew that buried within the *cytoarchitecture* of the cell were clues that would provide insight into the mysteries of life.

For a brief moment, the microscope's portholes became a crystal ball; in the eerie green glow of its fluorescent screen I saw my future. I knew I was going to be a cellular biologist whose research would focus on scrutinizing every nuance of the cell's ultrastructure to gain insights into the secrets of cellular life. As I had learned early on in graduate school, the *structure* and *function* of biological organisms are intimately intertwined. By correlating the cell's microscopic anatomy with its behavior, I was sure to gain insight into the nature of Nature. Throughout graduate school and postdoctoral research, and into my career as a medical school professor, my waking hours were consumed by explorations into the cell's molecular anatomy. For locked within the cell's structure were the secrets of its functions.

My exploration of the "secrets of life" led me into a research career studying the character of cloned human cells grown in tissue culture. Ten years after my first close encounter with an electron microscope, I was a tenured faculty member at the prestigious University of Wisconsin School of Medicine, internationally recognized for my research on cloned stem cells, and honored for my teaching skills. I had graduated to more powerful electron microscopes that allowed me to take three-dimensional CAT-scan-like rides through organisms where I had the opportunity to directly experience the molecular anatomy that provided for the magic of life. Though my tools were more sophisticated, my approach hadn't changed. I had never lost my seven-year-old conviction that the lives of the cells I studied had purpose.

Unfortunately, I had no such conviction that my own life had a purpose. I didn't believe in God, though I confess that on occasion I entertained the notion of a God who ruled with an extremely honed perverse sense of humor. I was after all a traditional biologist for whom God's existence is an unnecessary question: life is the consequence of blind chance, the flip of a friendly card, or, to be more precise, the random shake of genetic dice. The motto of our profession, since the time of Charles Darwin, has been: "God? We don't need no steenking God!"

It's not that Darwin denied the existence of God. He simply implied that chance, not Divine intervention, was responsible for the character of life on Earth. In his 1859 book, *The Origin of Species,* Darwin said that individual traits are passed from parents to their children. He suggested that "hereditary factors" passed from parent to child *control* the characteristics of an individual's life. That bit of insight set scientists off on a frenzied attempt to dissect life down to its molecular nuts and bolts, for within the structure of the cell was to be found the heredity mechanism that controlled life.

The search came to a remarkable end fifty years ago when James Watson and Francis Crick described the structure and function of the DNA double helix, the material of which genes are made. Scientists finally figured out the nature of the "hereditary factors" that Darwin had written about in the nineteenth century. The tabloids heralded the brave new world of genetic engineering with its promise of designer babies and magic bullet medical treatments. I vividly remember the large block print headlines that filled the front page on that memorable day in 1953: "Secret of Life Discovered."

Like the tabloids, biologists jumped on the gene bandwagon. The mechanism by which DNA controls biological life became the Central Dogma of molecular biology, painstakingly spelled out in textbooks. In the long-running debate over nature vs. nurture, the pendulum swung decidedly to nature. At first DNA was thought to be responsible only for our physical characteristics, but then we started believing that our genes control our emotions and behaviors as well. So if you are born with a defective happiness gene, you can expect to have an unhappy life.

Unfortunately, I thought I was one of those people victimized by a missing or mutant happiness gene. I was reeling from a relentless barrage of debilitating emotional roundhouse punches. My father had just died after a long, pain-fraught battle with cancer. I was his principal caretaker and had spent the previous four months flying back and forth between my job in Wisconsin and his home in New York every three or four days. In between stays at his deathbed, I was trying to maintain a research program, teach, and write a major grant renewal for the National Institutes of Health.

To further compound my stress levels, I was in the midst of an emotionally draining and economically devastating divorce. My financial resources were rapidly depleted as I tried to feed and clothe my new dependent, the judicial system. Economically challenged and homeless, I found myself living pretty much out of a suitcase in a most abysmal "garden" apartment complex. Most of my neighbors were hoping to upgrade their living standards by seeking accommodations in trailer parks. I was particularly scared of my next-door neighbors. My apartment was broken into, and my new stereo system was stolen in my first week of residence. A week later, six-foot tall, three-foot wide Bubba knocked on my door. Holding a quart of beer in one hand and picking his teeth with a ten-penny nail held in the other, Bubba wanted to know if I had the directions for the tape deck.

The nadir was the day I threw the phone through the glass door of my office, shattering the "Bruce H. Lipton, Ph.D., Associate Professor of Anatomy, U.W. School of Medicine" sign, all the while screaming, *"Get me out of here!"* My meltdown was precipitated by a phone call from a banker, who politely but firmly told me he couldn't approve my mortgage application. It was like the scene from *Terms of Endearment* when Debra Winger aptly responds to her husband's hopes for tenure: "We don't have enough money to pay the bills now. All tenure means is we won't have enough money forever!"

The Magic of Cells—Déjà Vu

Luckily, I found an escape in the form of a short-term sabbatical at a medical school in the Caribbean. I knew all my problems would not disappear there, but as the jet broke through the gray cloud cover above Chicago, it felt that way. I bit the inside of my cheek to prevent the smile on my face from evolving into audible laughter. I felt as joyful as my seven-year-old self, first discovering my life's passion, the magic of cells.

My mood lifted even more on the six-passenger commuter plane that took me to Montserrat, a mere four-by-twelve-mile dot in the Caribbean Sea. If there ever was a Garden of Eden, it probably would have resembled my new island home, erupting out of the sparkling aquamarine sea like a giant multifaceted emerald. When we landed, the gardenia-laced balmy breezes that swept the airport's tarmac were intoxicating.

The native custom was to dedicate the sunset period as a time of quiet contemplation, a custom I readily adopted. As each day wound down, I looked forward to the heavenly light show. My house, situated on a cliff fifty feet above the ocean, faced due west. A winding path through a tree-covered fern grotto led me down to the water. At the bottom of the grotto, an opening through a wall of jasmine bushes revealed a secluded beach, where I enhanced the sunset ritual by washing away the day with a few "laps" in the warm, gin-clear water. After my swim, I would mold the beach sand into a comfortable recliner, sit back, and watch the sun set slowly into the sea.

On that remote island, I was out of the rat race and free to see the world without the blinders of civilization's dogmatic beliefs. At first my mind was constantly reviewing and critiquing the debacle that was my life. But soon my mental Siskel and Ebert ceased their thumbs up/thumbs down review of my forty years. I began to re-experience what it was like to live in the moment and for the moment. To became reacquainted with sensations last experienced as a carefree child. To again *feel* the pleasure of being alive.

I became more human and more humane while living in that island paradise. I also became a better cell biologist. Almost all of my formal scientific training was in sterile, lifeless classrooms, lecture halls, and laboratories. However, once I was immersed in the Caribbean's rich ecosystem, I began to appreciate biology as a living, breathing, integrated system rather than a collection of individual species sharing a piece of the Earth's turf.

Sitting quietly within garden-like island jungles and snorkeling among the jeweled coral reefs gave me a window into the island's amazing integration of plant and animal species. All live in a delicate, dynamic balance, not only with other life forms but with the physical environment as well. It was life's harmony—not life's struggle—that sang out to me as I sat in the Caribbean Garden of Eden. I became convinced that contemporary biology pays too little attention to the important role of cooperation because its Darwinian roots emphasize life's competitive nature.

To the chagrin of my U.S. faculty colleagues, I returned to Wisconsin a screaming radical bent on challenging the sacred foundational beliefs of biology. I even began to openly criticize Charles Darwin and the wisdom of his theory of evolution. In the eyes of most other biologists, my behavior was tantamount to a priest bursting into the Vatican and claiming the Pope was a fraud.

My colleagues could be forgiven for thinking a coconut had hit me on the head when I quit my tenured position and, fulfilling my life's dream to be in a rock 'n' roll band, took off on a music tour. I discovered Yanni, who eventually became a big celebrity, and produced a laser show with him. But it soon became clear that I had a lot more aptitude for teaching and research than I did for producing rock 'n' roll shows. I wound down my midlife crisis, which I'll describe in more agonizing detail in a later chapter, by giving up the music business and returning to the Caribbean to teach cell biology again.

My final stop in conventional academia was at Stanford University's School of Medicine. By that time I was an unabashed proponent of a "new" biology. I had come to question not only Darwin's dog-eat-dog version of evolution but also biology's Central Dogma, the premise that genes control life. That scientific premise has one major flaw—genes cannot turn themselves on or off. In more scientific terms, genes are not "self-emergent." Something in the environment has to trigger gene activity. Though that fact had already been established by frontier science, conventional scientists blinded by genetic dogma had simply ignored it. My outspoken challenge of the Central Dogma turned me into even more of a scientific heretic. Not only was I a candidate for excommunication, I was now suitable for burning at the stake!

In a lecture during my interview at Stanford, I found myself accusing the gathered faculty, many of them internationally recognized geneticists, of being no better than religious fundamentalists for adhering to the Central Dogma despite evidence to the contrary. After my sacrilegious comments, the lecture room erupted into shouts of outrage that I thought meant the end of my job application. Instead, my insights concerning the mechanics of a new biology proved to be provocative enough to get me hired. With the support of some eminent scientists at Stanford, especially from the Pathology Department's chairman, Dr. Klaus Bensch, I was encouraged to pursue my ideas and apply them to research on cloned human cells. To the surprise of those around me, the experiments fully supported the alternative view of biology that I was postulating. I published two papers based on this research and left academia, this time for good. (Lipton, et al, 1991, 1992)

I left because, despite the support I got at Stanford, I felt that my message was falling on deaf ears. Since my departure, new research has consistently validated my skepticism about the Central Dogma and the primacy of DNA in controlling life. In fact, *epigenetics*, the study of the molecular mechanisms by which the environment controls gene activity, is today one of the most active areas of scientific research. The newly emphasized role of the environment in regulating gene activity was the focus of my cell research twenty-five years ago, long before the field of epigenetics was even established. (Lipton 1977a, 1977b) While that is gratifying for me intellectually, I know that if I were teaching and researching in a medical school, my colleagues would still be wondering about those coconuts because in the last decade I have become even more of a radical by academia's standards. My preoccupation with a new biology has become more than an intellectual exercise. I believe that cells teach us not only about the mechanisms of life, but also how to live rich, full lives.

In ivory tower science, that kind of thinking would no doubt win me the wacky Dr. Dolittle award for anthropomorphism or more precisely cytopomorphism—thinking like a cell—but for me it is Biology 101. You may consider yourself an individual, but as a cell biologist, I can tell you that you are in truth a cooperative community of approximately 50 trillion single-celled citizens. Almost all of the cells that make up your body are amoeba-like, individual organisms that have evolved a cooperative strategy for their mutual survival. Reduced to basic terms, human beings are simply the consequence of "collective amoebic consciousness." As a nation reflects the traits of its citizens, our human-ness must reflect the basic nature of our cellular communities.

Living the Lessons of Cells

Using these cell communities as role models, I came to the conclusion that we are not victims of our genes, but masters of our fates, able to create lives overflowing with peace, happiness, and love. I tested my hypothesis in my own life after a nudge from my audiences, who asked me why my insights hadn't made me any happier. They were right: I needed to integrate my new biological awareness into my daily life. I knew I had succeeded when, on a bright Sunday morning in the Big Easy, a coffee-shop waitress asked me: "Honey, you are the happiest person I ever did see. Tell me child, why are you so happy?" I was taken aback by her question, but nevertheless I blurted out, "I'm in Heaven!" The waitress shook her head from side to side mumbling, "My, my," and then proceeded to take my breakfast order. Well, it was true. I was happy, happier than I had ever been in my life.

A number of you critical readers may rightly be skeptical of my claim that Earth is Heaven. For by definition, Heaven is also the abode of the Deity and the blessed dead. Did I really think that New Orleans, or any other major city, could be part of Heaven? Ragged homeless women and children living in alleys; air so thick that one would never know if stars really existed; rivers and lakes so polluted that only unimaginable "scary" life forms could exist in them. This Earth is Heaven? The Deity lives here? He *knows* the Deity?

The answers to those questions are: yes, yes, and I believe I do. Well, to be completely honest, I must admit that I don't know all of the Deity personally, for I don't know all of you. For God's sake, there are over six billion of YOU. And to be more fully honest, I don't really know all of the members of the plant and animal kingdom either, though I believe they also comprise God.

In the immortal words of Tool Time's Tim Taylor: "Baaaaack the truck up! Is he saying that *humans* are God?"

Well . . . yes, I am. Of course I am not the first to have said that. It is written in Genesis that we are made in the image of God. Yes, this card-carrying rationalist is now quoting Jesus, Buddha, and Rumi. I have come full circle from a reductionist, scientific take on life to a spiritual one. We are made in the image of God, and we need to put Spirit back into the equation when we want to improve our physical and our mental health.

Because we are not powerless biochemical machines, popping a pill every time we are mentally or physically out of tune is not the answer. Drugs and surgery are powerful tools when they are not overused, but the notion of simple drug fixes is fundamentally flawed. Every time a drug is introduced into the body to correct function A, it inevitably throws off function B, C, or D. It is not gene-directed hormones and neurotransmitters that control our bodies and our minds; our beliefs control our bodies and our minds, and thus our lives . . . Oh ye of little belief!

The Light Outside of the Box

In this book I will draw the proverbial line in the sand. On one side of the line is a world defined by neo-Darwinism, which casts life as an unending war among battling, biochemical robots. On the other side of the line is the "New Biology," which casts life as a cooperative journey among powerful individuals who can program themselves to create joy-filled lives. When we cross that line and truly understand the New Biology, we will no longer fractiously debate the role of nurture and nature because we will realize that the fully conscious mind trumps both nature and nurture. And I believe we will also experience as profound a paradigmatic change to humanity as when a round-world reality was introduced to a flat-world civilization.

Humanities majors, who may be worried that this book offers an incomprehensible science lecture, have no fear. When I was an academic, I chafed at the three-piece, itchy suit, the constricting tie, the wing-tip shoes, and the interminable meetings, but I loved to teach. And in my post-academia life, I've gotten plenty of teaching practice; I have presented the principles of the New Biology to thousands of people all around the world. Through those lectures, I have honed my presentation of the science into easy-to-understand English illustrated by colorful charts, many of which are replicated in this book.

In Chapter 1, I discuss "smart" cells and why and how they can teach us so much about our own minds and bodies. In Chapter 2, I lay out the scientific evidence to show you genes do not control biology. I also introduce you to the latest discoveries of epigenetics, a booming field of biology that is unraveling the mysteries of how the environment influences the behavior of cells without changing the genetic code. It is a field that is uncovering new complexities in the nature of disease, including cancer and schizophrenia.

Chapter 3 is about the cell's membrane, the "skin" of the cell. You no doubt have heard more about the DNA-containing nucleus of the cell than you have about its membrane. But frontier science is revealing in ever greater detail what I concluded more than thirty years ago: that the membrane is the true brain of the cellular operation. And the latest research suggests that one day, this knowledge will lead to awesome medical breakthroughs.

In Chapter 4, I talk about the mind-bending discoveries of quantum physics. Those discoveries have profound implications for understanding and treating disease. Tragically, the conventional medical establishment has not yet incorporated quantum physics into its research or medical school training. (However, judging from my audiences, more and more insiders are hungry for new modalities.)

In Chapter 5, I explain why I named this book *The Biology of Belief*. Positive thoughts have a profound effect on behavior and genes, but *only* when they are in harmony with subconscious programming. And negative thoughts have an equally powerful effect. When we recognize how these positive and negative beliefs control our biology, we can use this knowledge to create lives filled with health and happiness.

Chapter 6 reveals why cells and people need to grow, how fear shuts down that growth, and how love, the opposite of fear, promotes growth.

Chapter 7 focuses on conscious parenting. As parents, we need to understand the role we play in programming our children's beliefs and the impact those beliefs have on our children's lives and thus the evolution of human civilization. This chapter is important even if you are not a parent, for as a former child, you'll find the insight into your own programming quite revealing!

In the Epilogue, I review how my understanding of the New Biology led me to realize the importance of integrating the realms of Spirit and Science, which was a radical shift from my background as an agnostic scientist. I am humbled to say that *Watkins Mind Body Spirit*, a magazine published by London's oldest esoteric bookshop, has named me one of the 100 Most Spiritually Influential Living People every year since it started the list in 2011. I am humbled that the list has put me in the same company as the Dalai Lama, Desmond Tutu, Wayne Dyer, Thich Nhat Hanh, Deepak

Chopra, Gregg Braden, and my publisher, Louise Hay, to name just a few. What an incredible honor for someone who used to study only the mechanistic, material world!

Are you ready to consider an alternate reality to that provided by the medical model—a reality in which the human body is not simply a biochemical machine? Are you ready to use your subconscious and conscious minds to create a life overflowing with health, happiness, and love without the aid of genetic engineers and without addicting yourself to drugs? There is nothing to buy, and there are no policies to take out. It is just a matter of temporarily suspending the archaic beliefs you have acquired from the scientific and media establishments so you can consider the exciting new awareness offered by leading-edge science.

CHAPTER 1

LESSONS FROM THE PETRI DISH:
In Praise of Smart Cells and Smart Students

On my second day in the Caribbean, as I stood in front of more than a hundred visibly on-edge medical students, I suddenly realized that not everyone viewed the island as a laid-back refuge. For these nervous students, Montserrat was not a peaceful escape but a last-ditch chance to realize their dreams of becoming doctors.

My class was geographically homogeneous, mostly American students from the East Coast, but there were all races and ages, including a sixty-seven-year-old retiree who was anxious to do more with his life. Their backgrounds were equally varied—former elementary school teachers, accountants, musicians, a nun, and even a drug smuggler.

Despite all the differences, the students shared two characteristics: One, they had failed to succeed in the highly competitive selection process that filled the limited number of positions in American medical schools. Two, they were "strivers" intent on becoming doctors—they were not about to be denied the opportunity to prove their qualifications. Most had spent their life savings or indentured themselves to cover the tuition and extra costs of living out of the country. Many found themselves completely alone for the first time in their lives, having left their families, friends, and loved ones behind. They put up with the most intolerable living conditions on that campus. Yet with all the drawbacks and the odds stacked against them, they were never deterred from their quest for a medical degree.

Well, at least that was true up to the time of our first class together. Prior to my arrival, the students had had three different histology/cell biology professors. The first lecturer left the students in the lurch when he responded to some personal issue by bolting from the island three weeks into the semester. In short order, the school found a suitable replacement who tried to pick up the pieces; unfortunately he bailed three weeks later because he got sick. For the preceding two weeks, a faculty member, responsible for another field of study, had been reading chapters out of a textbook to the class. This obviously bored the students to death, but the school was fulfilling a directive to provide a specified number of lecture hours for the course. Academic prerequisites set by American medical examiners have to be met in order for the school's graduates to practice in the States.

For the fourth time that semester, the weary students listened to a new professor. I briefed them on my background and my expectations for the course. I made it clear that even though we were in a foreign country, I was not going to expect any less from them than what was expected from my Wisconsin students. Nor should they want me to because to be certified all doctors have to pass the same Medical Boards, no matter where they go to medical school. Then I pulled a sheaf of exams out of my briefcase and told the students that I was giving them a self-assessment quiz. The middle of the semester had just passed, and I expected them to be familiar with half of the required course material. The test I handed out on that first day of the course consisted of twenty questions taken directly from the University of Wisconsin histology midterm exam.

The classroom was deadly silent for the first ten minutes of the testing period. Then nervous fidgeting felled the students one by one, faster than the spread of the deadly Ebola virus. By the time the twenty minutes allotted for the quiz were over, wide-eyed panic had gripped the class. When I said, "Stop," the pent-up nervous anxiety erupted into the din of a hundred excited conversations. I quieted the class down and began to read them the answers. The first five or six answers were met with subdued sighs. After I reached the tenth question, each subsequent answer was followed by agonizing groans. The highest score in the class was ten correct answers, followed by several students who answered seven correctly; with guesswork, most of the rest scored at least one or two correct answers.

When I looked up at the class, I was greeted with frozen, shell-shocked faces. The "strivers" found themselves behind the big eight ball. With more than half a semester behind them, they had to start the course all over again. A dark gloom overcame the students, most of whom were already treading water in their other, very demanding medical school courses. Within moments, their gloom had turned into quiet despair. In profound silence, I looked out over the students and they looked back at me. I experienced an internal ache—the class collectively resembled one of those Greenpeace pictures of wide-eyed baby seals just before heartless fur traders club them to death.

My heart welled. Perhaps the salt air and sweet scents had already made me more magnanimous. In any case, unexpectedly, I found myself announcing that I would make it my personal commitment to see that every student was fully prepared for the final exam, if they would commit to providing matching efforts. When they realized I was truly committed to their success, I could see the lights flash on in their previously panicked eyes.

Feeling like an embattled coach revving up the team for the Big Game, I told them I thought they were every bit as intelligent as the students I taught in the States. I told them I believed their stateside peers were simply more proficient at rote memorization, the quality that enabled them to score better in the medical college admissions tests. I also tried very hard to convince them that histology and cell biology are not intellectually difficult courses. I explained that in all of its elegance, nature employs very simple operating principles. Rather than just memorizing facts and figures, I promised they were going to gain an understanding of cells because I would present simple principles on top of simple principles. I offered to provide additional night lectures, which would tax their stamina after their already long lecture- and lab-packed days. The students were pumped up after my ten-minute pep talk. When the period ended, they bolted from that classroom snorting fire, determined they would not be beaten by the system.

After the students left, the enormity of the commitment I had made sank in. I started having doubts. I knew that a significant number of the students were truly unqualified to be attending medical school. Many others were capable students whose backgrounds had not prepared them for the challenge. I was afraid that my island idyll would degenerate into a frenetic, time-consuming academic scrimmage that would end in failure for my students and for me as their teacher. I started thinking about my job at Wisconsin, and suddenly it was beginning to look easy. At Wisconsin, I gave only eight lectures out of the approximately fifty that made up the histology/cell biology course. There were five members of the anatomy department who shared the lecturing load. Of course I was responsible for the material in all of the lectures because I was involved in their accompanying laboratory sessions. I was supposed to be available to answer all course-related questions asked by the students. But knowing the material and presenting lectures on the material are not the same thing!

I had a three-day weekend to wrestle with the situation I had created for myself. Had I faced a crisis such as this back home, my type A personality would have had me swinging from the proverbial chandeliers. Interestingly, as I sat by the pool, watching the sun set into the Caribbean, the potential angst simply morphed into an exciting adventure. I began to get excited about the fact that for the first time in my teaching career, I was solely responsible for this major course and free from having to conform to the style and content restrictions of team-taught programs.

Cells as Miniature Humans

As it turned out, that histology course was the most exhilarating and intellectually profound period of my academic career. Free to teach the course the way I wanted to teach it, I ventured into a new way of covering the material, an approach that had been roiling in my brain for several years. I had been fascinated by the idea that considering cells as "miniature humans" would make it easier to understand their physiology and behavior. As I contemplated a new structure for the course, I got excited. The idea of overlapping cell and human biology rekindled the inspiration for science I had felt as a child. I still experienced that enthusiasm in my research laboratory, though not when I was mired in the administrative details of being a tenured faculty member, including endless meetings and what, for me, were torturous faculty parties.

I was prone to thinking of cells as human-like because, after years behind a microscope, I had become humbled by the complexity and power of what at first appear to be anatomically simple, moving blobs in a petri dish. In school you may have learned the basic components of a cell: the nucleus that contains genetic material, the energy-producing mitochondria, the protective membrane at the outside rim, and the cytoplasm in between. But within these anatomically simple–looking cells is a complex world; these smart cells employ technologies that scientists have yet to fully fathom.

The notion of cells as miniature humans that I was mulling over would be considered heresy by most biologists. Trying to explain the nature of anything not human by relating it to human behavior is called anthropomorphism. "True" scientists consider anthropomorphism to be something of a mortal sin and ostracize scientists who knowingly employ it in their work.

However, I believed that I was breaking out of orthodoxy for a good reason. Biologists try to gain scientific understanding by observing nature and conjuring up a hypothesis of how things work. Then they design experiments to test their ideas. By necessity, deriving the hypothesis and designing the experiments require the scientist to "think" how a cell or another living organism carries out its life. Applying these "human" solutions, i.e., a human view of resolving biology's mysteries, automatically makes these scientists guilty of anthropomorphizing. No matter how you cut it, biological science is based to some degree on humanizing the subject matter.

Actually, I believe that the unwritten ban on anthropomorphism is an outmoded remnant of the Dark Ages, when religious authorities denied any direct relationship existed between humans and any of God's other creations. While I can see the value of the concept when people try to anthropomorphize a lightbulb, a radio, or a pocketknife, I do not see it as a valid criticism when it is applied to living organisms. Human beings are multicellular organisms—we must inherently share basic behavioral patterns with our own cells.

However, I know that it takes a shift in perception to acknowledge that parallel. Historically, our Judeo-Christian beliefs have led us to think that *we* are the intelligent creatures who were created in a separate and distinct process from all other plants and animals. This view has us looking down our noses at lesser creatures as nonintelligent life forms, especially those organisms on the lower evolutionary rungs of life.

Nothing could be further from the truth. When we observe other humans as individual entities or see ourselves in the mirror as an individual organism, in one sense, we are correct, at least from the perspective of our level of observation. However, if I brought you down to the size of an individual cell so you could see your body from that perspective, it would offer a whole new view of the world. When you looked back at yourself from that perspective you would not see yourself as a single entity. You would see yourself as a bustling community of more than 50 trillion individual cells.

As I toyed with these ideas for my histology class, the picture that kept recurring in my mind was a chart from an encyclopedia I had used as a child. Under the section on humans, there was an illustration with seven transparent plastic pages, each printed with an identical, overlapping outline of the human body. On the first page the outline was filled in with an image of a naked man. Turning the first page was like peeling off his skin and revealing his musculature, the image within the outline on the second page. When I turned the second page, the overlapping images of the remaining pages revealed a vivid dissection of the body. Flipping through the pages I could see in turn, the skeleton, the brain and nerves, blood vessels, and organ systems.

For my Caribbean course, I mentally updated those transparencies with several additional, overlapping pages, each illustrated with cellular structures. Most of the cell's structures are referred to as organelles, which are its "miniature organs" suspended within a jelly-like cytoplasm. Organelles are the functional equivalents of the tissues and organs of our own bodies. They include the nucleus, which is the largest organelle, the mitochondria, the Golgi body, and vacuoles. The traditional way of teaching the course is to deal first with these cellular structures, then move on to the tissues and organs of the human body. Instead, I integrated the two parts of the course to reflect the overlapping nature of humans and cells.

I taught my students that the biochemical mechanisms employed by cellular organelle systems are essentially the same mechanisms employed by our human organ systems. Even though humans are made up of trillions of cells, I stressed that there is not one "new" function in our bodies that is not already expressed in the single cell. Virtually every eukaryote (nucleus-containing cell) possesses the functional equivalent of our nervous system, digestive system, respiratory system, excretory system, endocrine system, muscle and skeletal systems, circulatory system, integument (skin), reproductive system, and even a primitive immune system, which utilizes a family of antibody-like "ubiquitin" proteins.

I also made it clear to my students that each cell is an intelligent being that can survive on its own, as scientists demonstrate when they remove individual cells from the body and grow them in a culture. As I knew intuitively when I was a child, these smart cells are imbued with intent and purpose; they actively seek environments that support their survival while simultaneously avoiding toxic or hostile ones. Like humans, single cells analyze thousands of stimuli from the microenvironment they inhabit. Through the analysis of this data, cells select appropriate behavioral responses to ensure their survival.

Single cells are also capable of learning through these environmental experiences and are able to create cellular memories, which they pass on to their offspring. For example, when a measles virus infects a child, an immature immune cell is called in to create a protective protein antibody against that virus. In the process, the cell must create a new gene to serve as a blueprint in manufacturing the measles antibody protein.

7

The first step in generating a specific measles antibody gene occurs in the nuclei of immature immune cells. Among their genes are a very large number of DNA segments that encode uniquely shaped snippets of proteins. By randomly assembling and recombining these DNA segments, immune cells create a vast array of different genes, each one providing for a uniquely shaped antibody protein. When an immature immune cell produces an antibody protein that is a "close" physical complement to the invading measles virus, that cell will be activated.

Activated cells employ an amazing mechanism called *affinity maturation* that enables the cell to perfectly "adjust" the final shape of its antibody protein, so that it will become a perfect complement to the invading measles virus. (Li, et al, 2003; Adams, et al, 2003) Using a process called *somatic hypermutation,* activated immune cells make hundreds of copies of their original antibody gene. However, each new version of the gene is slightly mutated so that it will encode a slightly different shaped antibody protein. The cell selects the variant gene that makes the best-fitting antibody. This selected version of the gene also goes through repeated rounds of somatic hypermutation to further sculpt the shape of the antibody to become a "perfect" physical complement of the measles virus. (Wu, et al, 2003; Blanden and Steele 1998; Diaz and Casali 2002; Gearhart 2002)

When the sculptured antibody locks on to the virus, it inactivates the invader and marks it for destruction, thus protecting the child from the ravages of measles. The cells retain the genetic "memory" of this antibody, so that in the future if the individual is again exposed to measles, the cells can immediately launch a protective immune response. The new antibody gene can also be passed on to all the cell's progeny when it divides. In this process, not only did the cell "learn" about the measles virus, it also created a "memory" that will be inherited and propagated by its daughter cells. This amazing feat of genetic engineering is profoundly important because it represents an inherent "intelligence" mechanism by which cells evolve. (Steele, et al, 1998)

The Origins of Life: Smart Cells Get Smarter

It shouldn't be surprising that cells are so smart. Single-celled organisms were the first life forms on this planet. Fossil evidence reveals they were here within 600 million years after the Earth was first formed. For the next 2.75 billion years of the Earth's history, only free-living, single-celled organisms—bacteria, algae, and amoeba-like protozoans—populated the world.

Around 750 million years ago, these smart cells figured out how to get smarter when the first multicellular organisms (plants and animals) appeared. Multicellular life forms were initially loose communities or "colonies" of single-celled organisms. At first, cellular communities consisted of from tens to hundreds of cells. But the evolutionary advantage of living in a community soon led to organizations comprised of millions, billions, and even trillions of socially interactive single cells. Though each individual cell is of microscopic dimensions, the size of multicellular communities may range from the barely visible to the monolithic. Biologists have classified these organized communities based on their structure as observed by the human eye. While the cellular communities appear as single entities to the naked eye—a mouse, a dog, a human—they are, in fact, highly organized associations of millions and trillions of cells.

The evolutionary push for ever-bigger communities is simply a reflection of the biological imperative to survive. The more awareness an organism has of its environment, the better its chances for survival. When cells band together they increase their awareness exponentially. If each cell were to be arbitrarily assigned an awareness value of X, then each colonial organism would collectively have a potential awareness value of at least X times the number of cells in the colony.

In order to survive at such high densities, the cells created structured environments. These sophisticated communities subdivided the workload with more precision and effectiveness than the ever-changing organizational charts that are a fact of life in big corporations. It proved more efficient for the community to have individual cells assigned to specialized tasks. In the development of animals and plants, cells begin to acquire these specialized functions in the embryo. A process of cytological specialization enables the cells to form the specific tissues and organs of the body. Over time, this pattern of *differentiation*, i.e., the distribution of the workload among the members of the community, became embedded in the genes of every cell in the community, significantly increasing the organism's efficiency and its ability to survive.

In larger organisms, for example, only a small percentage of cells are concerned with reading and responding to environmental stimuli. That is the role of groups of specialized cells that form the tissues and organs of the nervous system. The function of the nervous system is to perceive the environment and coordinate the behavior of all the other cells in the vast cellular community.

Division of labor among the cells in the community offered an additional survival advantage. The efficiency it offered enabled more cells to live on less. Consider the old adage: "Two can live as cheaply as one." Or consider the construction costs of building a two-bedroom single home versus the cost of building a two-bedroom apartment in a hundred-apartment complex. To survive, each cell is required to expend a certain amount of energy. The amount of energy conserved by individuals living in a community contributes to both an increased survival advantage and a better quality of life.

In American capitalism, Henry Ford saw the tactical advantage in the differentiated form of communal effort and employed it in creating his assembly line system of manufacturing cars. Before Ford, a small team of multiskilled workers would require a week or two to build a single automobile. Ford organized his shop so that every worker was responsible for only one specialized job. He stationed a large number of these differentiated workers along a single row, the assembly line, and passed the developing car from one specialist to the next. The efficiency of job specialization enabled Ford to produce a new automobile in ninety minutes rather than weeks.

Unfortunately, we conveniently "forgot" about the cooperation necessary for evolution when Charles Darwin emphasized a radically different theory about the emergence of life. He concluded 150 years ago that living organisms are perpetually embroiled in a "struggle for existence." For Darwin, struggle and violence are not only a part of animal (human) nature but the principal "forces" behind evolutionary advancement. In the final chapter of *The Origin of Species: By Means of Natural Selection, Or, the Preservation of Favoured Races in the Struggle for Life,* Darwin wrote of an inevitable "struggle for life" and that evolution was driven by "the war of nature, from famine and death." Couple that with Darwin's notion that evolution is random and you have a world, as poetically described by Tennyson, that can be characterized as "red in tooth and claw," a series of meaningless, bloody battles for survival.

Evolution Without the Bloody Claws

Though Darwin is by far the most famous evolutionist, the first scientist to establish evolution as a scientific fact was the distinguished French biologist Jean-Baptiste Lamarck. (Lamarck 1809, 1914, 1963) Even Ernst Mayr, the leading architect of "neo-Darwinism," a modernization of Darwin's theory that incorporates twentieth-century molecular genetics, concedes that Lamarck was the pioneer. In his classic 1970 book, *Evolution and the Diversity of Life,* (Mayr 1976, page 227) Mayr wrote: "It seems to me Lamarck has a much better claim to be designated the 'founder of the theory of evolution,' as indeed he has by several French historians . . . he was the first author to devote an entire book primarily to the presentation of a theory of organic evolution. He was the first to present the entire system of animals as a product of evolution."

Not only did Lamarck present his theory fifty years before Darwin, he offered a much less harsh theory of the mechanisms of evolution. Lamarck's theory suggested that evolution was based on an "instructive," cooperative interaction among organisms and their environment that enables life forms to survive and evolve in a dynamic world. His notion was that organisms acquire and pass on adaptations necessary for their survival in a changing environment. Interestingly, Lamarck's hypothesis about the mechanisms of evolution conform to modern cell biologists' understanding of how immune systems adapt to their environment as described above.

Lamarck's theory was an early target of the Church. The notion that humans evolved from lower life forms was denounced as heresy. Lamarck was also scorned by his fellow scientists who, as creationists, ridiculed his theories. A German developmental biologist, August Weismann, helped propel Lamarck into obscurity when he tried to test Lamarck's theory that organisms pass on survival-oriented traits acquired through their interaction with the environment. In one of Weismann's experiments, he cut off the tails of male and female mice and mated them. Weismann argued that if Lamarck's theory were correct, the parents should pass on their tail-less state to future generations. The first generation of mice was born with tails. Weismann repeated the experiment for twenty-one more generations, but not one tail-less mouse was born, leading Weismann to conclude that Lamarck's notion of inheritance was wrong.

But Weismann's experiment was not a true test of Lamarck's theory. Lamarck suggested that such evolutionary changes could take "immense periods of time," according to biographer L. J. Jordanova. In 1984, Jordanova wrote that Lamarck's theory "rested on" a number of "propositions" including "the laws governing living things have produced increasingly complex forms over immense periods of time." (Jordanova 1984, page 71) Weismann's five-year experiment was clearly not long enough to test the theory. An even more fundamental flaw in his experiment is that Lamarck never argued that every change an organism experienced would take hold. Lamarck said organisms hang on to traits (like tails) when they need them to survive. Although Weismann didn't think the mice needed their tails, no one asked the mice if they thought their tails were necessary for survival!

Despite its obvious flaws, the study of the tail-less mice helped destroy Lamarck's reputation. In fact, Lamarck has been mostly ignored or vilified. Cornell University evolutionist C. H. Waddington wrote in *The Evolution of an Evolutionist* (Waddington 1975, page 38): "Lamarck is the only major figure in the history of biology whose name has become to all intents and purposes, a term of abuse. Most scientists' contributions are fated to be outgrown, but very few authors have written works, which, two centuries later, are still rejected with indignation so intense that the skeptic may suspect something akin to an uneasy conscience. In point of fact, Lamarck has, I think, been somewhat unfairly judged."

Waddington wrote those prescient words thirty-five years ago. Today Lamarck's theories are being re-evaluated under the weight of a body of new science that suggests that the oft-denounced biologist was not entirely wrong and the oft-lauded Darwin not entirely correct. The title of an article in the prestigious journal *Science* in 2000 was one sign of glasnost: "Was Lamarck Just a Little Bit Right?" (Balter 2000)

One reason some scientists are taking another look at Lamarck is that evolutionists are reminding us of the invaluable role cooperation plays in sustaining life in the biosphere. Scientists have long noted symbiotic relationships in nature. In *Darwin's Blind Spot* (Ryan 2002, page 16), British physician Frank Ryan chronicles a number of such relationships, including a yellow shrimp that gathers food while its partner gobi fish protects it from predators and a species of hermit crab that carries a pink anemone on top of its shell. "Fish and octopuses like to feed on hermit crabs, but when they approach this species, the anemone shoots out its brilliantly colored tentacles, with their microscopic batteries of poisoned darts, and stings the potential predator, encouraging it to look elsewhere for its meal." The warrior anemone gets something out of the relationship as well because it eats the crab's leftover food.

But today's understanding of cooperation in nature goes much deeper than the easily observable relationships. "Biologists are becoming increasingly aware that animals have coevolved and continue to coexist, with diverse assemblages of microorganisms that are required for normal health and development," according to a recent article in *Science* called "We Get By with a Little Help from Our (Little) Friends." (Ruby, et al, 2004) The study of these relationships is now a rapidly growing field called "Systems Biology."

Ironically, in recent decades, we have been taught to wage war against microorganisms with everything from antibacterial soap to antibiotics. But that simplistic message ignores the fact that many bacteria are essential to our health. The classic example of how humans get help from microorganisms is the bacteria in our digestive system, which are essential to our survival. The bacteria in our stomach and intestinal tract help digest food and also enable the absorption of life-sustaining vitamins. This microbe-human cooperation is the reason that the rampant use of antibiotics is detrimental to our survival. Antibiotics are indiscriminate killers; they kill bacteria that are required for our survival as efficiently as they kill harmful bacteria.

Recent advances in genome science have revealed an additional mechanism of cooperation among species. Living organisms, it turns out, actually integrate their cellular communities by sharing their genes. It had been thought that genes are passed on only to the progeny of an individual organism through reproduction. Now scientists realize that genes are shared not only among the individual members of a species but also among members of different species. The sharing of genetic information via *gene transfer* speeds up evolution since organisms can acquire "learned" experiences from other organisms. (Nitz, et al, 2004; Pennisi 2004; Boucher, et al, 2003; Dutta and Pan 2002; Gogarten 2003) Given this sharing of genes, organisms can no longer be seen as disconnected entities; there is no wall between species. Daniel Drell, manager of the Department of Energy's microbial genome program told *Science* (2001 294:1634) "we can no longer comfortably say what is a species anymore." (Pennisi 2001)

This sharing of information is not an accident. It is nature's method of enhancing the survival of the biosphere. As discussed earlier, genes are physical memories of an organism's learned experiences. The recently recognized exchange of genes among individuals disperses those memories, thereby influencing the survival of all organisms that make up the community of life. Now that we are aware of this inter- and intra-species gene transfer mechanism, the dangers of genetic engineering become apparent. For example, tinkering with the genes of a tomato may not stop at that tomato but could alter the entire biosphere in ways that we cannot foresee. Already there is a study that shows that when humans digest genetically modified foods, the artificially created genes transfer into and alter the character of the beneficial bacteria in the intestine. (Heritage 2004; Netherwood, et al, 2004) Similarly, gene transfer among genetically engineered agricultural crops and surrounding native species has given rise to highly resistant species deemed superweeds. (Milius 2003; Haygood, et al, 2003; Desplanque, et al, 2002; Spencer and Snow 2001) Genetic engineers have never taken the reality of gene transfer into consideration when they have introduced genetically modified organisms into the environment. We are now beginning to experience the dire consequences of this oversight as their engineered genes are spreading among and altering other organisms in the environment. (Watrud, et al, 2004; Biello 2010)

Genetic evolutionists warn that if we fail to apply the lessons of our shared genetic destiny, which should be teaching us the importance of cooperation among all species, we threaten human existence. We need to move beyond Darwinian Theory, which stresses the importance of *individuals*, to one that stresses the importance of the *community*. British scientist Timothy Lenton provides evidence that evolution is more dependent on the interaction among species than it is on the interaction of individuals within a species. Evolution becomes a matter of the survival of the fittest *groups* rather than the survival of the fittest individuals. In a 1998 article in *Nature*, Lenton wrote that rather than focusing on individuals and their role in evolution "we must consider the totality of organisms and their material environment to fully understand which traits come to persist and dominate." (Lenton 1998)

Lenton subscribes to James Lovelock's Gaia hypothesis that holds that the Earth and all of its species constitute one interactive, living organism. Those who endorse this hypothesis argue that tampering with the balance of the superorganism called Gaia, whether it be by destroying the rainforest, depleting the ozone layer, or altering organisms through genetic engineering, can threaten its survival and consequently ours.

Recent studies funded by Britain's Natural Environment Research Council provide support for those concerns. (Thomas, et al, 2004; Stevens, et al, 2004) While there have been five mass extinctions in the history of our planet, they are all presumed to have been caused by extraterrestrial events, such as a comet smashing to Earth. One of the new studies concludes that the "natural world is experiencing the sixth, major extinction event in its history." (Lovell 2004) This time though, the cause of the extinction is not extraterrestrial. According to one of the study's authors, Jeremy Thomas, "As far as we can tell this one is caused by one animal organism—man."

Walking the Talk of Cells

In my years of teaching in medical school, I had come to realize that medical students in an academic setting are more competitive and backbiting than a truckload of lawyers. They live out the Darwinian struggle in their quest to be one of the "fittest" who stagger to graduation after four grueling years in medical school. The single-minded pursuit of stellar medical school grades, without regard for the students surrounding you, no doubt follows a Darwinian model, but it always seemed to me an ironic pursuit for those who are striving to become compassionate healers.

But my stereotypes about medical students toppled during my stay on the island. After my call to arms, my class of misfits stopped acting like conventional medical students; they dropped their survival of the fittest mentality and amalgamated into a single force, a team that helped them survive the semester. The stronger students helped the weaker and, in so doing, all became stronger. Their harmony was both surprising and beautiful to observe.

In the end, there was a bonus: a happy Hollywood ending. For their final exam, I gave my students exactly the same test the students in Wisconsin had to pass. There was virtually no difference in the performance of these "rejects" and their "elite" counterparts in the States. Many students later reported that when they went home and met with their peers who attended American medical schools, they proudly found themselves more proficient in their understanding of the principles governing the life of cells and organisms.

I was of course thrilled that my students had pulled off an academic miracle. But it was years before I understood *how* they were able to do it. At the time, I thought the format of the course was key, and I still believe that overlapping human and cell biology is a better way to present the course material. But now that I've ventured into what I told you would be considered by some as wacky Dr. Dolittle territory, I think a good part of the reason for my students' success was that they eschewed the behavior of their counterparts in the United States. Instead of mirroring smart American medical students, they mirrored the behavior of smart cells, banding together to become even smarter. I didn't tell my students to pattern their lives after the lives of the cells, because I was still steeped in traditional, scientific training. But I like to think that they went in that direction intuitively after listening to my praise of cells' ability to group together cooperatively to form more complex and highly successful organisms.

I didn't know it at the time, but I now believe that another reason for my students' success was that I did not stop at praising cells. I praised the students as well. They needed to hear they were first-rate students in order to believe that they could perform as first-rate students. As I will detail in future chapters, so many of us are leading limited lives not because we have to but because we *think* we have to. But I'm getting ahead of myself. Suffice it to say that after four months in paradise, teaching in a way that clarified my thinking about cells and the lessons they provide to humans, I was well on my way to an understanding of the New Biology, which leaves in the dust the defeatism of genetic and parental programming as well as survival-of-the-fittest Darwinism.

❋ ❋ ❋

When I first wrote this chapter, I had to search hard for the first glimmerings that the much-maligned Jean-Baptiste Lamarck would finally be credited for his insights about evolution. Nevertheless, proverbial optimist that I am, as you read above, I included a reference to an article with the tentative headline, "Was Lamarck Just a Little Bit Right?" I'm happy to report that my optimism was warranted. A decade later, it's a lot easier to find Lamarck supporters who believe that he was more than "just a little bit" right, that, in fact, he was a seer!

Nearly 200 years after his death, epigenetic research, one of the hottest fields in science today, is corroborating over and over Lamarck's oft-ridiculed belief that organisms adapt to their environment and can pass on those adaptations to future generations. Consider this definitive (no question mark!) headline I quickly came across during my research for this anniversary edition: "The Rebirth of Lamarckism (The Rise of Epigenetics)." (Rogers 2009)

Of course, Lamarck did not have any insight into the molecular nature of genes and their relationship to organismal expression (neither did Darwin), so I can't argue that he was actually an epigeneticist. It has taken the high-tech labs of modern researchers to uncover the subtle chemical modifications to DNA and DNA-associated proteins that enable organisms to adapt to their environment and pass on those adaptations to their offspring without changing the structure of DNA molecules. Lamarck's theory of the inheritance of acquired characteristics, cited as the primary reason to debunk Lamarck, has now been found to be a valid hereditary mechanism. (Morris 2012) Frontier research is not only helping rehabilitate Lamarck's reputation, it is also undermining our culture's belief in genetic determinism, which, as you know by now, is one of the major themes of *The Biology of Belief*—the genes we inherit from our mothers and our fathers are not our fate!

I don't want to oversell the scientific community's shift to Lamarckism. When it comes to the mechanisms that drive evolution, there is still a lot of debate. For example, when the theory of "adaptive mutation," which holds that mutations occur in response to specific stresses, was first brought to academic attention in the 1980s by eminent physician and molecular biologist Dr. John Cairns, he was called a heretic, and this theory is still controversial today. (Cairns, et al, 1988) Adaptive mutation conflicts with neo-Darwinism's focus on chance alterations in heredity based on *natural selection*, a process that was described by Darwin as the "struggle for life most severe" and that came to be known as "survival of the fittest." (Though it's a catchy phrase, survival of the fittest is actually a tautology, an obvious truth that is not an apt way of describing the driving forces of evolution. By definition, *fittest* means "most capable of survival," so the phrase can be rewritten as "survival of the most capable of surviving." No argument there!)

Neo-Darwinism attributes mutations to accidental copying mistakes in replicating the genes; if the genetic error enhances the organism's survivability, the mutation is selected to propagate. This suggests that the direction of evolutionary advancement is accidental and unpredictable . . . how's that for a tautology! In response to the perennial questions "How did we get here?" and "Why are we here?" neo-Darwinian theory would lead us to believe we evolved through a few billion years of "lucky" genetic accidents. In contrast, Lamarckian theory implies that evolution-producing mutations arise from an organism's "need" to adapt to life-threatening environmental stresses, so they are not random and to a large degree are environmentally predictable.

This seemingly arcane scientific debate is important because adaptive mutations imply purposefulness in biological evolution—the purpose being to conform to prevailing conditions in the surrounding environment, which includes the entire community of life. Eventually, I believe the theory of adaptive mutations will prevail and provide more support for the view that the web of life and the process of evolution are the result of a highly organized, symbiotic *collaboration* among all living organisms.

The fascinating research of biologist and mathematician Martin A. Nowak, Director of Harvard's Program for Evolutionary Dynamics, already provides support for the crucial role of cooperation in evolution. Using mathematical and computer simulations, Nowak divided populations into "cooperators," those who support others, and "defectors," those who do not support others even after accepting help from others. Nowak found that in the several thousand papers scientists have published on how cooperators, ranging from bacteria to human beings, prevail in evolution, all the scenarios fall into five categories. (Nowak 2012)

One category, for example, is "spatial selection," in which cooperators and defectors are not uniformly distributed in a population. In these populations with "patches of cooperators," helpful individuals band together and prevail against defectors. Another category is what Nowak calls the "I'll scratch your back, and someone will scratch mine," in which an individual decides to be a cooperator because of the person in need's reputation. He uses the example of Japanese macaques: low-ranking monkeys that groom high-ranking ones may improve their reputations (and receive more grooming) by being seen with the high-ranking monkeys Nowak calls "the top brass."

Nowak found that cooperation-defection works on several levels—an individual can simultaneously be a cooperator and a defector. The example Nowak uses is a group of employees at a company who compete ruthlessly against one another for promotions but also cooperate with one another to ensure that their company beats the performance of other companies. That insight about the complex nature of cooperation-defection is in alignment with the principles of systems biology—another field that has boomed in the last decade—which recognizes that biological insights emerge best from studying the dynamics of interacting systems rather than focusing on only one system. One case in point: medical science once attempted to understand heart disease by focusing on the function and structure of the heart. However, fundamental breakthroughs in cardiac disease were only recognized when the heart's function was studied in relation to the influence of other systems, such as the nervous, neuroendocrine, immune, and digestive systems.

Nowak's models also confirm what everyone who is agonizing over the current dismal state of our planet has noted—that cooperation is "intrinsically unstable": there are cycles when defection prevails. However, he also offers the good news that "the altruistic spirit always seems to rebuild itself." Nowak's sums up what he has discovered through his simulations, with the conclusion that "life is not just a struggle for survival but also a snuggle for survival."

Now more than ever, we need more research on the cooperative snuggle for survival lest we fall into a defection cycle during which we destroy ourselves and our planet. I believe we have been brought to the brink by our misunderstanding of evolution as simply a continuous struggle and quest for individual fitness (as measured by the number of one's "toys"). Human civilization has bought into the warning couched in the subtitle of Darwin's *Origin of Species* book: *The Preservation of Favoured Races in the Struggle for Life*—in other words, that life is an all-out struggle wherein the riches go to the fittest, regardless of the means by which they are attained.

According to this "scientific" principle, the less fit genetically deserve only what's left over . . . if anything. That mentality has brought us continuous wars over material possessions, overconsumption that has led to unsustainable resource exploitation, and increasingly unequal wealth distribution as well as an obviously ailing planet. The Darwinian focus on the fitness of the individual de-emphasizes the significance of communal cooperation in evolution.

One of the most striking areas where we have ignored the importance of cooperation among organisms is in our own bodies. In the decade since I decried our "war against microorganisms with everything from antibacterial soap to antibiotics," a wealth of damaging evidence has emerged about the toll this war is taking on our bodies.

The fact is that hundreds of trillions of microbial "invaders," mostly in our gut, are absolutely necessary for our survival, and there are ten times more of them than cells in the human body. Because the body cannot survive without its microbes (collectively called the "microbiome"), they are the functional equivalent of any of our other vital organ systems. In (belated) recognition of the importance of the microbiome, humans and most other organisms are now properly defined as superorganisms (complex organisms composed of many smaller organisms). (Saey 2013A) Again in belated recognition of the microbiome's importance, in 2007, the National Institutes of Health created the Human Microbiome Project to study it. Those scientists reported that humans and other animals form a life-sustaining bond with their gut microbes. Researchers have found that human genes influence the genetics of the microbiome, and the microbiome's genes (that make up 99 percent of the unique genes in our body!) regulate genes in our cells. (Saey 2013B)

In his alarming new book, *Missing Microbes: How the Overuse of Antibiotics Is Fueling Our Modern Plagues*, Dr. Martin J. Blaser, Director of the Human Microbiome Program at New York University, warns not only about antibiotic resistance but also about the declining diversity of the human microbiome that is increasing our susceptibility to chronic conditions from allergies and asthma to diabetes and obesity. For example, type 1 diabetes has been doubling in incidence about every twenty years in the industrialized world; in Finland, the incidence has risen 550 percent since 1950. Blaser writes that these modern epidemics are "not only diseases but also external signs of internal change." Recent studies have found that "otherwise normal individuals have lost 15 to 40 percent of their microbial diversity and the genes that accompany it" mostly due to the overprescription of broad-spectrum antibiotics that kill microbes indiscriminately. Yet Blaser, who has studied the microbes that populate our bodies for thirty years, calls them and their 20 million genes the "guerrilla warriors" that help us fight disease. (Blaser 2014)

While Blaser is warning about the declining diversity of our microbiome, other scientists are pointing with alarm to the declining diversity of our planet, where animal populations and species are decreasing at an alarming rate. Stanford scientists have tracked species abundance and population numbers over a period of time and found that extinction rates are up to a thousand times higher than they would be if people weren't in the environment generating pollution, deforesting, monocropping, and overharvesting. (Dirzo, et al, 2014) Many environmental scientists believe we have crossed the threshold for a major environmental collapse and are in the throes of the sixth mass extinction event to hit this planet.

Environmentalists have long known that the structure of localized ecological systems can shift abruptly and irreversibly from one state to another when stressed to critical thresholds. Evidence now indicates that the entire global ecosystem can react in the same abrupt way and is, in fact, currently in danger of doing so. Anthony Barnosky, a professor at the University of California, Berkeley's Department of Integrative Biology, and others argue that we are at a planetary "tipping point" because human activities are inducing Mother Earth to express a critical global transition. (Barnosky, et al, 2012) A recent study by NASA confirms that global industrial civilization is heading toward collapse in coming decades (i.e., soon!). (Ahmed 2014)

Civilization did not create global climate change (the planet has already been through five ice ages), but our behavior and technology are generating environmental stressors that exacerbate the impact of the climate change crisis. The process of societal rise-and-collapse has been a cyclical phenomenon throughout history, and in some cases, those collapse periods have lasted for centuries. While previous collapses primarily impacted localized human social systems, the coming collapse has already had a profound global impact on the health of the planet.

We now live in an era known as the Anthropocene, which empha-sizes that human activities are causing massive changes to our natural world at an unprecedented rate. Not one location on our planet, from the southern tip of Antarctica to the heights of Mt. Everest, has re-mained untouched by human influence. For example, fossil fuel burn-ing has left an imprint on our immediate environment while the thin veil of the Earth's atmosphere carries it to all portions of the globe. This reminds us of the following: (1) that we are all connected; (2) that we all leave an imprint; and (3) that the Earth that sustains us is finite. Today's global crises are warnings that we must stop exploiting the abundance and vitality of our living home and begin to reconnect and honor the planet as many traditional societies have done for eons.

Well, that's a cheery picture! However, as a flagrant optimist, I pre-fer to consider the positive side of Nature's resiliency. In 1883, a series of eruptions on Krakatoa in Indonesia led to new volcanic islands aris-ing out of the sea. Lava flows on one of the islands in 1960 eliminated all life forms and left the island in a condition scientists actually refer to as a state of "sterilization." Surveys and studies monitoring the rise of flora and fauna on the islands for over five decades documented the abundance of an incredibly diverse ecology that has been thriving on these "sterile" islands since that time. In the aftermath of its cata-strophic disturbance, the island's vital and thriving ecological paradise has since become more robust, expressing a diverse plasticity that en-hances its ability to resist environmental stress. (Whittaker, et al, 1989) This lesson from Nature emphasizes the old adage, "What doesn't kill you will make you stronger."

I also take heart from the fact that organismal cooperation is not a nagging exception to the rule of evolution but instead one of its primary architects and that humans are (though it's hard to believe sometimes!), in Nowak's words, "supercooperators." Collectively, the cooperative accomplishments of human civilization have taken us to the Moon and beyond, and I hope our collective accomplish-ments will also take us to a restored planet, a restored microbiome, and beyond. After all, I have personally seen the dramatically positive changes that can occur when the cooperative behavior among my Caribbean medical students helped them evolve to become better humans, and more importantly, compassionate healers.

CHAPTER 2

IT'S the ENVIRONMENT, STUPID

I will never forget a piece of wisdom I received in 1967, on the first day I learned to clone stem cells in graduate school. It took me decades to realize how profound this seemingly simple piece of wisdom was for my work and my life. My professor, mentor, and consummate scientist Irv Konigsberg was one of the first cell biologists to master the art of cloning stem cells. He told me that when the cultured cells you are studying are ailing, you look first to the cell's environment, not to the cell itself, for the cause.

My professor wasn't as blunt as Bill Clinton's campaign manager, James Carville, who decreed, "It's the economy, stupid," to be the mantra for the 1992 presidential election. But cell biologists would have done well to post, "It's the environment, stupid," over our desks, just as the "It's the economy, stupid" sign was posted at Clinton headquarters. Though it wasn't apparent at the time, I eventually realized that this advice was a key insight into understanding the nature of life. Over and over I learned the wisdom of Irv's advice. When I provided a healthy environment for my cells, they thrived; when the environment was less than optimal, the cells faltered. When I adjusted the environment, these "sick" cells revitalized.

But most cell biologists knew nothing of this wisdom of tissue culture techniques. And scientists moved sharply away from considering environmental influences after Watson and Crick's revelation of DNA's genetic code. Even Charles Darwin conceded, near the end of his life, that his evolutionary theory had shortchanged the role of the environment. In an 1876 letter to Moritz Wagner he wrote: "In my opinion, the greatest error which I have committed has been not allowing sufficient weight to the direct action of the environments, i.e., food, climate, etc., independently of natural selection . . . When I wrote the *Origin,* and for some years afterwards, I could find little good evidence of the direct action of the environment; now there is a large body of evidence." (Darwin, F 1888)

Unfortunately, Darwin's followers perceived that his return to Lamarckian "thinking" was a sign of Darwin's aging and now addled mind. Rather than following their master's revised vision, Darwinian evolutionists chose to remain more Darwinian than Darwin! The problem with the Darwinian underemphasis on the environment is that it led to an overemphasis on "nature" in the form of genetic determinism—the belief that genes "control" biology. This belief has not only led to a misallocation of research dollars, as I will argue in a later chapter, but, more importantly, it has changed the way we think about our lives. When you are convinced that genes control your life and you know that you had no say in which genes you were saddled with at conception, you have a good excuse to consider yourself a victim of heredity. "Don't blame me for my work habits—it's not my fault that I've been procrastinating on my deadline . . . It's genetic!"

Since the dawning of the Age of Genetics, we have been programmed to accept that we are subservient to the power of our genes. The world is filled with people who live in constant fear that, on some unsuspecting day, their genes are going to turn on them. Consider the masses of people who think they are ticking time bombs; they wait for cancer to explode in their lives as it exploded in the life of their mother or brother or sister or aunt or uncle. Millions of others attribute their failing health not to a combination of mental, physical, emotional, and spiritual causes but simply to the inadequacies of their body's biochemical mechanics. Are your kids unruly? Increasingly the first choice is to medicate these children to correct their "chemical imbalances" rather than fully grappling with what is going on in their bodies, minds, and spirits.

Of course there is no doubt that some diseases, like Huntington's chorea, beta thalassemia, and cystic fibrosis, can be blamed entirely on one faulty gene. But single-gene disorders affect less than 2 percent of the population; the vast majority of people come into this world with genes that should enable them to live a happy and healthy life. The diseases that are today's scourges—diabetes, heart disease, and cancer—short circuit a happy and healthy life. These diseases, however, are not the result of a single gene, but of complex interactions among multiple genes and environmental factors.

What about all those headlines trumpeting the discovery of a gene for everything from depression to schizophrenia? Read those articles closely and you'll see that behind the breathless headline is a more sober truth. Scientists have linked lots of genes to lots of different diseases and traits, but scientists have rarely found that *one* gene causes a trait or a disease. In the realm of human diseases, defective genes acting alone only account for about 2 percent of our total disease load. (Strohman 2003)

The confusion occurs when the media repeatedly distort the meaning of two words: correlation and causation. It's one thing to be linked to a disease; it's quite another to cause a disease, which implies a directing, controlling action. If I show you my keys and say that a particular key "controls" my car, you at first might think that makes sense because you know you need that key to turn on the ignition. But does the key actually "control" the car? If it did, you couldn't leave the key in the car alone because it might just borrow your car for a joy ride when you are not paying attention. In truth, the key is "correlated" with the control of the car; the person who turns the key actually controls the car. Specific genes are correlated with an organism's behavior and characteristics. But these genes are not activated until something triggers them.

What activates genes? The answer was elegantly spelled out in 1990 in a paper entitled *Metaphors and the Role of Genes and Development* by H. F. Nijhout. (Nijhout 1990) Nijhout presents evidence that the notion that genes control biology has been so frequently repeated for such a long period of time that scientists have forgotten it is a hypothesis, not a truth. In reality, the idea that genes control biology is a supposition, which has never been proven and, in fact, has been undermined by the latest scientific research. Genetic control, argues Nijhout, has become a metaphor in our society. We want to believe that genetic engineers are the new medical magicians who can cure diseases and while they're at it create more Einsteins and Mozarts as well. But metaphor does not equate with scientific truth. Nijhout summarizes the truth: "When a gene product is needed, a signal from its environment, not an emergent property of the gene itself, activates expression of that gene." In other words, when it comes to genetic control, "It's the environment, stupid."

Protein: The Stuff of Life

It is easy to understand how genetic control became a metaphor as scientists with ever-greater excitement zeroed in on the mechanisms of DNA. Organic chemists discovered that cells are made up of four types of very large molecules: polysaccharides (complex sugars), lipids (fats), nucleic acids (DNA/RNA), and proteins. Though the cell requires each of the four molecular types, proteins are the most important single component for living organisms. Our cells are, in the main, an assembly of protein building blocks. So one way of looking at our trillion-celled bodies is that they are protein machines, although, as you know, I think we are more than machines! It sounds simple, but it isn't. For one thing, it takes over 100,000 different types of proteins to run our bodies.

Let's take a closer look at how our cells' ~100,000 proteins are assembled. Each protein is a linear string of linked amino acid molecules, comparable to a child's pop bead necklace, as illustrated at the top of the following page.

Each bead represents one of the twenty amino acid molecules used by cells. Though I like the pop bead analogy because everyone is familiar with it, it is not an exact one because each amino acid has a slightly different shape. So to be completely accurate, you should think of a pop bead necklace that got mangled a bit in the factory.

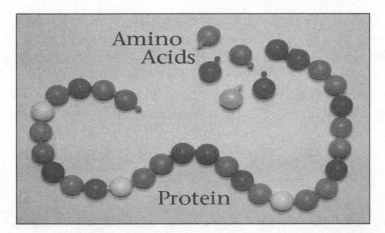

And to be even more accurate, you should know that the amino acid necklace, which forms the "backbone" of the cells' proteins, is far more malleable than a pop bead necklace, which falls apart when you bend it too much. The structure and behavior of the linked amino acids in the protein backbones better resemble that of a snake's backbone, as shown below. (©Warren Jacobi/Corbis) The spine of a snake, made up of a large number of linked subunits, the vertebrae, is capable of coiling the snake into a wide variety of shapes, ranging from a straight rod to a knotted "ball."

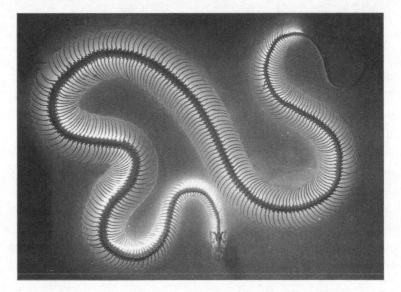

The flexible links *(peptide bonds)* between amino acids in a protein backbone enable each protein to adopt a variety of shapes. Through the rotation and flexion of their amino acid "vertebrae," protein molecules resemble nano-snakes in their ability to writhe and squirm. There are two primary factors that determine the contour of a protein's backbone and therefore its shape. One factor is the physical pattern defined by the sequence of differently shaped amino acids comprising the pop-bead-like backbone.

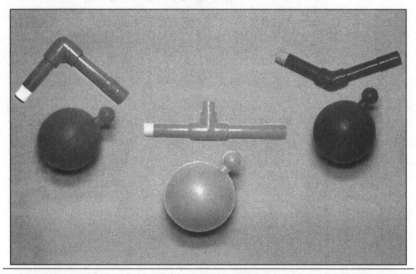

Unlike uniform-shaped pop beads, each of the twenty amino acids comprising protein backbones has a unique shape (conformation). Consider the differences between the character of a "backbone" made from identically shaped pop beads and one assembled from the differently shaped pipe fittings illustrated above.

The second factor concerns the interaction of electromagnetic charges among the linked amino acids. Most amino acids have positive or negative charges, which act like magnets: *like* charges cause the molecules to repel one another, while *opposite* charges cause the molecules to attract each other. As shown on the following page, a protein's flexible backbone spontaneously folds into a preferred shape when its amino acid subunits rotate and flex their bonds to balance the forces generated by their positive and negative charges.

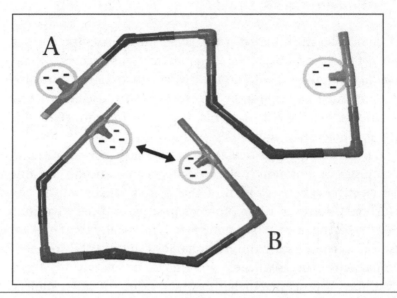

The protein backbones shown in A and B have the exact same amino acid (pipe fitting) sequence but reveal radically different conformations. Variations in the backbone's shape result from differential rotations at the junctions between adjacent pipe fittings. Like the pipe fittings illustrated above, the protein's differently shaped amino acids also rotate around their junctions (peptide bonds), allowing the backbone to wriggle like a snake. Proteins shape-shift though they will generally prefer two or three specific conformations. Which of the two conformations, A or B, would our hypothetical protein prefer? The answer is related to the fact that the two terminal amino acids (pipe fittings) have regions of negative charges. Since like charges repel each other, the farther apart they are, the more stable the conformation. Conformation A would be preferred because the negative charges are farther apart than they are in B.

The backbones of some protein molecules are so long that they require the assistance of special "helper" proteins called chaperones to aid in the folding process. Improperly folded proteins, like people with spinal defects, are unable to function optimally. Such aberrant proteins are marked for destruction by the cell; their backbone amino acids are disassembled and recycled in the synthesis of new proteins.

How Proteins Create Life

Living organisms are distinguished from nonliving entities by the fact that they move; they are *animated*. Cells harness the energy of protein movements to do the "work" that characterizes living systems, such as respiration, digestion, and muscle contraction. To understand the nature of life, one must first understand how protein "machines" are empowered to move.

The final shape, or *conformation* (the technical term used by biologists), of a protein molecule reflects a balanced state among the electromagnetic charges of the amino acids comprising the backbone. However, if the protein's positive and negative charges are altered, the protein backbone will dynamically twist and adjust itself to accommodate the new charges. The distribution of electromagnetic charges within a protein can be selectively altered by a number of processes including the binding of other molecules or chemical groups such as hormones, the enzymatic removal or addition of charged atoms (ions) in the backbone's amino acids, or interference from electromagnetic fields such as those emanating from cell phones. (Tsong 1989)

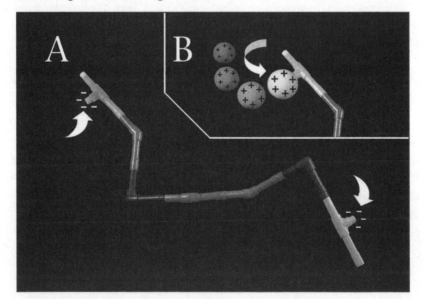

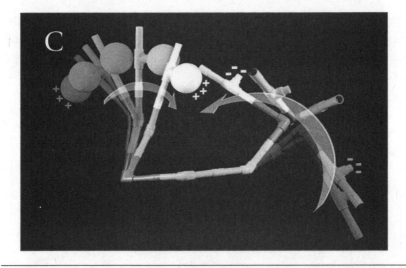

Figure A shows the preferred conformation of our hypothetical protein backbone. The repelling forces between the two negatively charged terminal amino acids (arrows) causes the backbone to extend so that the negative amino acids are as far apart as possible. Figure B shows a close-up of an end amino acid. A signal, in this case a molecule with a very positive electric charge (white sphere), is attracted to and binds with the negative site on the protein's terminal amino acid. In our particular scenario, the signal is more positive in charge than the amino acid is negative in charge. After the signal couples with the protein, there is now an excess positive charge at this end of the backbone. Since positive and negative charges attract one another, the backbone's amino acids will rotate around their bonds so that positive and negative terminals will come closer together. Figure C shows the protein changing from conformation A to conformation B. Changing conformations generates movement and the movement is harnessed to do work, providing for such functions as digestion, respiration, and muscle contraction. When the signal molecule detaches, the protein returns to its preferred extended conformation. This is how signal-generated protein movements provide for life.

The shape-shifting proteins exemplify an even more impressive engineering feat because their precise, three-dimensional shapes also give them the ability to link up with other proteins. When a protein encounters a molecule that is a physical and energetic complement, the two bind together like human-made products with interlocking gears, say an eggbeater or an old-fashioned watch.

Examine the following two illustrations. The first shows five uniquely shaped proteins, examples of the molecular "gears" found in cells. These organic "gears" have softer edges than machine-shop-manufactured gears, but you can see that their precise, three-dimensional shapes would enable them to securely engage with other complementary proteins.

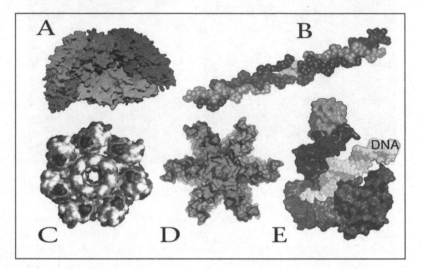

Protein Menagerie. Illustrated above are five different examples of protein molecules. Each protein possesses a precise three-dimensional conformation that is the same for each copy of itself in every cell. A) Enzyme that digests hydrogen atoms; B) Woven filament of collagen protein; C) Channel, a membrane-bound protein with hollow central pore; D) Protein subunit of "capsule" that encloses a virus; E) DNA-synthesizing enzyme with attached helical DNA molecule

In the second illustration (p. 35), I chose a wind-up watch to represent the workings of the cell. The first picture shows a metal machine, revealing the gears, springs, jewels, and case of the watch model. When Gear A turns it causes Gear B to turn. When B moves it causes Gear C to turn, etc. In the next image, I overlay the human-made machine gears with softer-edged organic proteins (magnified millions of times in proportion to the watch) so that it becomes visually conceivable that proteins could be like the watch's mechanism. In this metal-protein "machine," one can imagine Protein A rotating and causing Protein B to revolve, which in turn causes Protein C to move. Once you see that possibility, you can look to the third figure in which the human-made parts are removed. Voilà! We are left with a protein "machine," one of the thousands of similar protein assemblies that collectively comprise the cell!

Cytoplasmic proteins that cooperate in creating specific physiologic functions are grouped into specific assemblies known as *pathways*. These assemblies are identified by the functions they perform, such as respiration pathways, digestion pathways, muscle contraction pathways, and the infamous, energy-generating Krebs cycle, the bane of many a science student who has to memorize every one of its protein components and complex chemical reactions.

Can you imagine how excited cell biologists were when they figured out how the protein machines work? Cells exploit the movements of these protein assembly machines to empower specific metabolic and behavioral functions. The constant, shape-shifting movements of proteins—

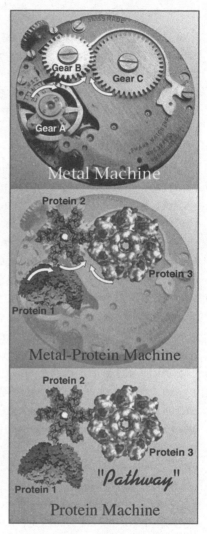

Metal Machine

Metal-Protein Machine

"*Pathway*"

Protein Machine

which can occur thousands of times in a single second—are the movements that propel life.

The Primacy of DNA

You'll notice that, in the above section, I didn't discuss DNA at all. That's because it is the changing of the proteins' electromagnetic charges that is responsible for their behavior-generating movement, not DNA. How did we get to the widespread and often-cited notion that genes "control" biology? In the *Origin of Species,* Darwin suggested that "hereditary" factors were passed on from generation to generation, controlling the traits of the offspring. Darwin's influence was so great that scientists myopically focused on identifying that hereditary material, which, they thought, controlled life.

In 1910, intensive microscopic analyses revealed that the hereditary information passed on generation after generation was contained in chromosomes, thread-like structures that become visible in the cell just before it divides into two "daughter" cells. Chromosomes are incorporated into the daughter cell's largest organelle, the nucleus. When scientists isolated the nucleus, they dissected the chromosomes and found that the hereditary elements were essentially comprised of only two kinds of molecules, protein and DNA. Somehow the protein machinery of life was entangled in the structure and function of these chromosome molecules.

The understanding of the chromosome's functions was further refined in 1944 when scientists determined that it was DNA that actually contained hereditary information. (Avery, et al, 1944; Lederberg 1994) The experiments that singled out DNA were elegant. These scientists isolated pure DNA from one species of bacteria—let's call it Species A—and added the pure DNA to cultures containing only Species B bacteria. Within a short time, Species B bacteria began to show hereditary traits that were formerly seen only in Species A. Once it was known that you needed nothing other than DNA to pass on traits, the DNA molecule became a scientific superstar.

It was now left to Watson and Crick to unravel the structure and function of that superstar molecule. DNA molecules are long and thread-like. They are made from four nitrogen-containing chemicals called bases (adenine, thymine, cytosine, and guanine, abbreviated as A, T, C, and G). Watson and Crick's discovery of DNA's structure led to the fact that the sequence of the A, T, C, and G bases in DNA spells out the sequence of amino acids along a protein's backbone (Watson and Crick 1953). Those long strings of DNA molecules can be subdivided into single genes, segments that provide the blueprint for specific protein backbones. The code for recreating the protein machinery of the cell had been cracked!

Watson and Crick also explained why DNA is the perfect hereditary molecule. Each DNA strand is normally intertwined with a second strand of DNA, a loosely wrapped configuration known as the "double helix." The genius of this system is that the sequences of DNA bases on both strands are mirror images of each other. When the two strands of DNA unwind, each single strand contains the information to make an exact, complementary copy of itself. So through a process of separating the strands of a double helix, DNA molecules become self-replicating. This observation led to the assumption that DNA "controlled" its own replication . . . it was its own "boss."

The "suggestion" that DNA controlled its own replication *and* served as the blueprint for the body's proteins led Francis Crick to create biology's Central Dogma, the belief that DNA rules. The dogma was so fundamental to modern biology it was essentially written in stone, the equivalent of science's Ten Commandments. The dogma, also referred to as "the Primacy of DNA," is a fixture of almost every scientific text.

In the dogma's scheme of how life unfolds, DNA perches loftily on top, followed by RNA. RNA is the short-lived Xerox copy of the DNA. As such, it is the physical template encoding the amino acid sequence that makes up a protein's backbone. The Primacy of DNA diagram provides the logic for the Age of Genetic Determinism. Because the character of a living organism is defined by the nature of its proteins and its proteins are encoded in the DNA, then by logic, DNA would represent the "first cause," or primary determinant of an organism's traits.

The Central Dogma's assumption of a one-way flow of information from DNA to RNA to protein is profoundly important. Since proteins represent the physical body, the dogma implies that your physical body, and your life experiences cannot send information back and alter the DNA. According to the Dogma, DNA controls your life and you cannot influence your DNA!

The Human Genome Project

After DNA achieved superstar status, the remaining challenge was to create a catalog of all the genetic stars in the human firmament. Enter the Human Genome Project, a global scientific effort begun in the late 1980s to create a catalog of all the genes present in humans.

From the outset, the Human Genome Project was a massively ambitious one. Conventional thought held that the body needed one gene to provide the blueprint for each of the 100,000-plus different proteins that make up our bodies. Add to that at least 20,000 regulatory genes, which orchestrate the activity of the protein-encoding genes. Scientists concluded that the human genome would contain a minimum of 120,000 genes located within the twenty-three pairs of human chromosomes.

But that wasn't the whole story. A cosmic joke was unfolding, one of those jokes that periodically unsettle scientists convinced they have discovered the secrets of the universe. Consider the impact of Nicolaus Copernicus' discovery published in 1543 that the Earth was not the center of the universe, as was thought by the scientist-theologians of the day. The fact that the Earth actually revolved around the sun and that the sun itself was not the center of the universe undermined the teachings of the Church. Copernicus' paradigm-busting discoveries launched the modern, scientific revolution by challenging the presumed "infallibility" of the Church. Science eventually displaced the Church as Western civilization's source of wisdom for understanding the mysteries of the universe.

Geneticists experienced a comparable shock when, contrary to their expectations of over 120,000 genes, they found that the entire human genome consists of fewer than 25,000 genes. (Pennisi 2003a and 2003b; Pearson 2003; Goodman 2003) Over 80 percent of the presumed and *required* DNA does not exist! The missing genes proved to be more troublesome than the missing eighteen minutes of the Nixon tapes. The one-gene, one-protein concept was a fundamental tenet of genetic determinism. Now that the Human Genome Project has toppled the one-gene for one-protein concept, our current theories of how life works have to be scrapped. No longer is it possible to believe that genetic engineers can, with relative ease, fix all our biological dilemmas. There are simply not enough genes to account for the complexity of human life or of human disease.

The Central Dogma. The dogma, also referred to as the Primacy of DNA, defines the flow of information in biological organisms. As indicated by the arrows, the flow is only in one direction, from DNA to RNA and then to protein. The DNA represents the cell's long-term memory, passed from generation to generation. RNA, an unstable copy of the DNA molecule, is the active memory that is used by the cell as a physical template in synthesizing proteins. Proteins are the molecular building blocks that provide for the cell's structure and behavior. DNA is implicated as the "source" that controls the character of the cell's proteins, hence the concept of DNA's primacy that literally means "first cause."

I may sound like Chicken Little shouting that the genetics sky is falling. However, you don't have to take my word for it. Chicken Big said the same thing. In a commentary on the surprising results of the Human Genome Project, David Baltimore, one of the world's preeminent geneticists and a Nobel Prize winner, addressed the issue of human complexity (Baltimore 2001): "But unless the human genome contains a lot of genes that are opaque to our computers, it is clear that we do not gain our undoubted complexity over worms and plants by using more genes.

"Understanding what does give us our complexity—our enormous behavioral repertoire, ability to produce conscious action, remarkable physical coordination, precisely tuned alterations in response to external variations of the environments, learning, memory, need I go on?—remains a challenge for the future."

As Baltimore states, the results of the Human Genome Project force us to consider other ideas about how life is controlled. "Understanding what does give us our complexity . . . remains a challenge for the future." The sky *is* falling.

In addition, the results of the Human Genome Project are forcing us to reconsider our genetic relationship with other organisms in the biosphere. We can no longer use genes to explain why humans are at the top of the evolutionary ladder. It turns out there is not much difference in the total number of genes found in humans and those found in primitive organisms. Let's take a look at three of the most studied animal models in genetic research, a microscopic nematode roundworm known as *Caenorhabditis elegans*, the fruit fly, and the laboratory mouse.

The primitive *Caenorhabditis* worm serves as a perfect model for studying the role of genes in development and behavior. This rapidly growing and reproducing organism has a precisely patterned body comprised of exactly 969 cells and a simple brain of about 302 cells. Nonetheless it has a unique repertoire of behaviors and, most importantly, it is amenable to genetic experimentation. The *Caenorhabditis* genome consists of approximately 24,000 genes. (Blaxter 2003) The human body, comprised of over 50 trillion cells, contains only about 1,000 more genes than the lowly, spineless, thousand-celled microscopic worm.

The fruit fly, another favored research subject, has 15,000 genes. (Blaxter 2003; Celniker, et al, 2002) So the profoundly more complicated fruit fly has 9,000 fewer genes than the more primitive *Caenorhabditis* worm. And when it comes to the question of mice and men, we might have to think more highly of them or less of ourselves; the results of parallel genome projects reveal that humans and rodents have roughly the same number of genes!

Cell Biology 101

In retrospect, scientists should have known that genes couldn't provide the *control* of our lives. By definition, the brain is the organ responsible for controlling and coordinating the physiology and behavior of an organism. Conventional science, as revealed in a recent publication by the U.S. Department of Health and Human Services (2005), perceives that the nucleus is "basically the cell's brain": "It contains the equivalent of the cell's gray matter—its genetic material, or DNA. In the form of genes, each with a host of helper molecules, DNA determines the cell's identity, masterminds its activities and is the official cookbook for the body's proteins."

Since genes were presumed to "control" the traits of the cell and the nucleus is the organelle that contains virtually all the cell's DNA, considering the nucleus as the "brain" of the cell made sense.

But is the nucleus truly the cell's brain? If our assumption that the nucleus and its DNA-containing material is the "brain" of the cell, then removing the cell's nucleus, a procedure called enucleation, should result in the immediate death of the cell.

And now, for the big experiment . . . (Maestro, a drumroll if you please).

The scientist drags our unwilling cell into the microscopic operating arena and straps it down. Using a micromanipulator, the scientist guides a needle-like micropipette into position above the cell. With a deft thrust of the manipulator, our investigator plunges the pipette deep into the cell's cytoplasmic interior. By applying a little suction, the nucleus is drawn up into the pipette, and the pipette is withdrawn from the cell. Below the nucleus-engorged pipette lies our sacrificial cell—its "brain" torn out.

But *wait!* It's still moving! My God . . . the cell is still *alive!*

The wound has closed and like a recovering surgical patient, the cell begins to slowly stagger about. Soon the cell is back on its feet (okay, its pseudopods), fleeing the microscope's field with the hope that it will never see a doctor again.

Following enucleation, many cells can survive for up to two or more months without genes. Viable enucleated cells do not lie about like brain-dead lumps of cytoplasm on life-support systems. These cells actively ingest and metabolize food, maintain coordinated operation of their physiologic systems (respiration, digestion, excretion, motility, etc.), retain an ability to communicate with other cells, and are able to engage in appropriate responses to growth and protection requiring environmental stimuli.

Unsurprisingly, enucleation is not without side effects. Without their genes, cells are not able to divide, nor are they able to reproduce any protein parts they lose through the normal wear and tear of the cytoplasm. The inability to replace defective cytoplasmic proteins contributes to mechanical dysfunctions that ultimately result in the death of the cell.

Our experiment was designed to test the idea that the nucleus is the "brain" of the cell. If the cell had died immediately following enucleation, the observations would have at least supported that belief. However, the results are unambiguous: enucleated cells still exhibit complex, coordinated, life-sustaining behaviors, which imply that the cell's "brain" is still intact and functioning.

The fact that enucleated cells retain their biological functions in the absence of genes is by no means a new discovery. Over a hundred years ago, classical embryologists routinely removed the nuclei from dividing egg cells and showed that a single, enucleated egg cell was able to develop as far as the blastula, an embryonic stage consisting of forty or more cells. Today, enucleated cells are used for industrial purposes as living "feeder" layers in cell cultures designed for virus vaccine production.

If the nucleus and its genes are not the cell's brain, then what exactly is DNA's contribution to cellular life? Enucleated cells die, not because they have lost their brain but because they have lost their reproductive capabilities. Without the ability to reproduce their parts, enucleated cells cannot replace failed protein building blocks, nor replicate themselves. So the nucleus is not the brain of the cell—the nucleus is the cell's gonad! Confusing the gonad with the brain is an understandable error because science has always been and still is a patriarchal endeavor. Males have often been accused of thinking with their gonads, so it's not entirely surprising that science has inadvertently confused the nucleus with the cell's brain!

Epigenetics: The New Science of Self-Empowerment

Genes-as-destiny theorists have obviously ignored hundred-year-old science about enucleated cells, but they cannot ignore new research that undermines their belief in genetic determinism. While the Human Genome Project was making headlines, a group of scientists were inaugurating a new, revolutionary field in biology called *epigenetics.* The science of epigenetics, which literally means "control above genetics," profoundly changes our understanding of how life is controlled. (Pray 2004; Silverman 2004) In the last decade, epigenetic research has established that DNA blueprints passed down through genes are not set in concrete at birth. Genes are not destiny! Environmental influences, including nutrition, stress, and emotions, can modify those genes without changing their basic blueprint. And those modifications, epigeneticists have discovered, can be passed on to future generations as surely as DNA blueprints are passed on via the double helix. (Reik and Walter 2001; Surani 2001; Watters 2006; Cloud 2010)

There is no doubt that epigenetic discoveries have lagged behind genetic discoveries. Since the late 1940s, biologists have been isolating DNA from the cell's nucleus in order to study genetic mechanisms. In the process they extract the nucleus from the cell, break open its enveloping membrane, and remove the chromosomal contents, half of which is made up of DNA and half of which is made up of regulatory proteins. In their zeal to study DNA, most scientists threw away the proteins, which we now know is the equivalent of throwing the baby out with the bathwater. Epigeneticists are bringing back the baby, by studying the chromosome's proteins, and those proteins are turning out to play as crucial a role in heredity as DNA.

In the chromosome, the DNA forms the core, and the proteins cover the DNA like a sleeve. When the genes are covered, their information cannot be "read." Imagine your bare arm as a piece of DNA representing the gene that codes for blue eyes. In the nucleus, this stretch of DNA is covered by bound regulatory proteins, which cover your blue-eye gene like a shirtsleeve, making it impossible to be read.

Primacy of Environment. The new science reveals that the information that controls biology starts with environmental signals that, in turn, control the activity of regulatory proteins on the DNA. Regulatory proteins direct the activity of genes. The DNA, RNA, and protein functions are the same as described in the Primacy of DNA chart. Note: the flow of information is no longer unidirectional. In the 1960s, Howard Temin challenged the Central Dogma with experiments that revealed RNA could go against the predicted flow of information and rewrite the DNA program. Originally ridiculed for his "heresy," Temin later won a Nobel Prize for describing reverse transcriptase, the molecular mechanism by which RNA can rewrite the genetic code. Reverse transcriptase is now notorious, for it is used by the AIDS virus' RNA to commandeer the infected cell's DNA. It is also now known that epigenetic changes in the DNA molecule, such as adding or removing methyl chemical groups, influence the binding of regulatory proteins. Proteins must also be able to buck the predicted flow of information, since protein antibodies in immune cells are involved with changing the DNA in the cells that synthesize them. The size of the arrows indicating information flow are intentionally not the same. There are tight restrictions on the reverse flow of information, a design that would prevent radical changes to the cell's genome.

How do you get that sleeve off? You need an environmental signal to spur the "sleeve" protein to change shape, i.e., detach from the DNA's double helix, allowing the gene to be read. Once the DNA is uncovered, the cell makes a copy of the exposed gene. As a result, the activity of the gene is "controlled" by the presence or absence of the ensleeving proteins, which are in turn controlled by environmental signals.

The story of epigenetic control is the story of how environmental signals control the activity of genes. It is now clear that the Primacy of DNA chart described earlier is outmoded. The revised scheme of information flow should now be called the "Primacy of the Environment." The new, more sophisticated flow of information in biology starts with an environmental signal, then goes to a regulatory protein and only then goes to DNA, RNA, and the end result, a protein.

The science of epigenetics has also made it clear that there are two mechanisms by which organisms pass on hereditary information. Those two mechanisms provide a way for scientists to study both the contribution of nature (genes) and the contribution of nurture (epigenetic mechanisms) in human behavior. If you only focus on the DNA blueprints, as scientists have been doing for decades, the influence of the environment is impossible to fathom. (Dennis 2003; Chakravarti and Little 2003)

Let's present an analogy that hopefully will make the relationship between epigenetic and genetic mechanisms clearer. Are you old enough to remember the days when television programming stopped after midnight? After the normal programming signed off, a "test pattern" would appear on the screen. Most test patterns looked like a dartboard with a bull's eye in the middle, similar to the one pictured on the following page.

Think of the pattern of the test screen as the pattern encoded by a given gene, say the one for brown eyes. The dials and switches of the TV fine-tune the test screen by allowing you to turn it on and off and modulate a number of characteristics, including volume, color, hue, contrast, brightness, and vertical and horizontal holds. By adjusting the dials, you can alter the appearance of the pattern on the screen, while not actually changing the original broadcast pattern. This is precisely the role of regulatory proteins. Studies of protein synthesis reveal that epigenetic "dials" can create 2,000 or more variations of proteins from the same gene blueprint. (Bray 2003; Schmuker, et al, 2000)

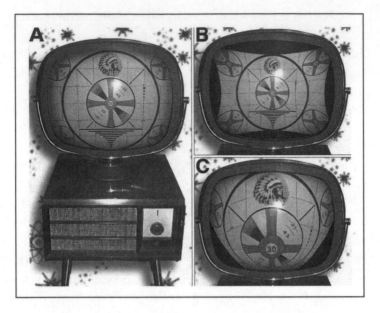

In this epigenetic analogy, the test pattern on the screen represents the genetic code (program). While the TV's controls can change the appearance of the pattern (B and C), they do not change the original pattern of the broadcast (i.e., the gene). Epigenetic control modifies the readout of a gene without changing the DNA code.

Parental Life Experiences Shape Their Children's Genetic Character

We now know that the environmentally influenced fine-tuning described above can be passed from generation to generation. A landmark Duke University study published in the August 1, 2003 issue of *Molecular and Cellular Biology* found that an enriched environment can even override genetic mutations in mice. (Waterland and Jirtle 2003) In the study, scientists looked at the effect of dietary supplements on pregnant mice with the abnormal "agouti" gene. Agouti mice have yellow coats and are extremely obese, which predisposes them to cardiovascular disease, diabetes, and cancer.

Agouti Sisters. One-year-old female genetically identical agouti mice. Maternal methyl donor supplementation shifts coat color of the offspring from yellow to brown and reduces the incidence of obesity, diabetes, and cancer. (Photo courtesy of Jirtle and Waterland©)

In the experiment, one group of yellow, obese, agouti mothers received methyl-group-rich supplements available in health food stores: folic acid, vitamin B12, betaine, and choline. Methyl-rich supplements were chosen because a number of studies have shown that the methyl chemical group is involved with epigenetic modifications. When methyl groups attach to a gene's DNA, it changes the way regulatory chromosomal proteins bind to the DNA molecule. If the proteins bind too tightly to the gene, the protein sleeve cannot be removed and the gene cannot be read. Methylating DNA can silence or modify gene activity.

This time the headlines "Diet Trumps Genes" were accurate. The mothers who got the methyl-group-rich supplements produced standard, lean, brown mice, even though their offspring had the same agouti gene as their mothers. The agouti mothers who didn't get the supplements produced yellow pups, which ate much more than the brown pups. The yellow pups wound up weighing almost twice as much as their lean, "pseudo-agouti" counterparts.

The University's photo, shown above, is striking. Though the two mice are genetically identical, they are radically different in appearance: one mouse is lean and brown and the other mouse is obese and yellow. What you can't see in the picture is that the obese mouse is diabetic while its genetically identical counterpart is healthy.

Other studies have found epigenetic mechanisms to be a factor in a variety of diseases, including cancer, cardiovascular disease, and diabetes. In fact, only 5 percent of cancer and cardiovascular patients can attribute their disease directly to heredity. (Willett 2002; Silverman 2004) While the media made a big hoopla over the discovery of the BRCA1 and BRCA2 breast cancer genes, they failed to emphasize that 95 percent of breast cancers are not due to inherited genes. The malignancies in a significant number of cancer patients are derived from environmentally induced epigenetic alterations and not defective genes. (Kling 2003; Jones 2001; Seppa 2000; Baylin 1997) Recently, eminent scientist and physician Dean Ornish revealed that by just changing diet and lifestyle for ninety days, prostate cancer patients switched the activity of over 500 genes. Many of their gene changes inhibited biological processes critical in the formation of their tumors. (Ornish, et al, 2008)

The epigenetic evidence has become so compelling that some brave scientists are even invoking the "L" word for Jean-Baptiste Lamarck, the much-scorned evolutionist, who believed that traits acquired as a result of environmental influence could be passed on. Philosopher Eva Jablonka and biologist Marion Lamb wrote in their 1995 book *Epigenetic Inheritance and Evolution—The Lamarckian Dimension:* "In recent years, molecular biology has shown that the genome is far more fluid and responsive to the environment than previously supposed. It has also shown that information can be transmitted to descendants in ways other than through the base sequence (code) of DNA." (Jablonka and Lamb 1995; Kaiser 2005)

We're back to where we started in this chapter, the environment. In my own work in the laboratory, I saw over and over the impact a changed environment had on the cells I was studying. But it was only at the end of my research career, at Stanford, that the message fully sank in. I saw that endothelial cells, which are the blood vessel–lining cells I was studying, changed their structure and function depending on their environment. When, for example, I added inflammatory chemicals to the tissue culture, the endothelial cells rapidly became the equivalent of macrophages, the scavengers of the immune system. What was also exciting to me was that the cells transformed even when I destroyed their DNA with gamma rays. These endothelial cells were "functionally enucleated," yet they completely changed their biological behavior in response to inflammatory agents, just as they had when their nuclei were intact. These cells were clearly showing some "intelligent" control in the absence of their genes. (Lipton 1991; Butler, et al, 2010)

Twenty years after my mentor Irv Konigsberg's advice to first consider the environment when your cells are ailing, I finally got it. DNA does not control biology, and the nucleus itself is not the brain of the cell. Just like you and me, cells are shaped by where they live. In other words, it's the environment, stupid.

✳ ✳ ✳

The exploding field of epigenetic research has not only made Jean-Baptiste Lamarck look like a seer, it has made my professor and mentor Irv Konigsberg, who inspired the title for this chapter, look like more of one as well. More than forty years later, it's still the environment, stupid!

Consider a Stanford study touted in the media with headlines that sound like *The Biology of Belief*! (I'll try to restrain myself from pointing out over and over that the newest research supports the conclusions of the first edition of *The Biology of Belief*, though that's hard for me because I've felt so many times like a voice in the wilderness.) From *U.S. News*: "Environment Trumps Genes at Shaping Immune System: Study." (Preidt 2015) From *ScienceDaily*: "Environment, not genes, dictates human immune variation, study finds." (Goldman 2015)

The Stanford study found that three quarters of the variations in the immune systems of identical twins (who share the same genome) were due to "nonheritable," environmental influences including exposure to microbes, toxins, diet, and vaccinations. The study found that environmental factors over time shaped each twin's immune system, with the result that the differences in identical twins over sixty are greater than the differences in twins under twenty. (Brodin, et al, 2015) Mark Davis, Stanford microbiologist and immunologist and lead researcher, says, "A healthy human immune system continually adapts to its encounters with hostile pathogens, friendly gut microbes, nutritional components and more, overshadowing the influences of most heritable factors." (Goldman 2015)

It's become ever more obvious that the belief that sequencing someone's genome could predict what diseases they would succumb to later in life is false. As for the Human Genome Project, the giant monkey wrench that it threw into the conventional perception of evolution has gotten even more giant. When I first wrote this chapter, the current research led me to give humans a thousand-gene advantage over the simple *Caenorhabditis* worm, but now, even that small-number advantage has disappeared. Recent technical advances in reading the genome have further reduced the number of genes found in humans to only about 19,000, the *same* number now estimated for the *Caenorhabditis* worm. In fact, by now, the origin of over 90 percent of human genes has been traced back to more than a hundred million years ago, which implies that worm and human genomes likely share most of the same genes. (Ezkurdia, et al, 2014, Madhusoodanan 2014)

So in terms of a gene-based metric to score evolution, we humans have been hurtled even more definitively down to the base of German embryologist Ernst Haeckel's "Tree of Life," an illustration he created in 1886 shortly after Lamarck and then Darwin introduced the science of evolution. Haeckel's image traced the lineage of animal evolution from the simplest (bacteria) at the trunk of the tree to humans occupying the tree's top branches. That lineage made sense when the primacy of DNA was held by science as the controlling factor of life—evolution biologists naturally assumed that, as one ascended the Tree, higher evolutionary traits would result in greater genetic complexity. But with humans now at the bottom of a gene-based Tree, it is even more evident that gene populations do not determine organismal evolution.

Here's another fact I often use in my lectures as a cautionary tale about overemphasizing the roles of genes: the same gene used to code for the protein keratin found in hair also provides for *all* of the following structures: skin, nails, claws, hooves, and horns. The gene that encodes the synthesis of individual keratin protein building blocks *does not control* how those keratin molecules will be used.

So protein-encoding genes do provide for cellular building blocks, but do not determine an organism's structure or its complexity. That leaves us with a fundamental question: what does?

Figuring out an answer to that question leads us to another completely unexpected outcome of the Genome Project: genes that encode a cell's protein building blocks constitute less than 2 percent of the genome's total amount of DNA, so the vast majority of DNA does *not* contribute to the cell's protein population. The belief that this DNA lacked function led Francis Crick to label it as "junk DNA." That term, though readily accepted by the public, irritates large numbers of biologists who cannot fathom the idea that cells carry massive amounts of "useless" DNA. That's why geneticists prefer to use the term "dark matter" when referring to noncoding DNA.

Bent on unlocking the mysteries of this dark matter, a consortium of genetic scientists created the ENCODE Project (an acronym for the *Enc*yclopedia *of DNA E*lements) to assess the function of the genome's so-called junk DNA. Their research to date, published after I wrote the first edition of this book, reveals that over 80 percent of *noncoding* DNA is involved with regulating the production and assembly of gene-encoded proteins. A major discovery also found that "dark" DNA contains mechanisms by which *environmental* information can be used to modify the readout of protein-encoding genes. It turns out that dark DNA uses epigenetic mechanisms that enable a human cell with 19,000 gene blueprints to code for over a hundred thousand different protein molecules! (Ecker 2012)

Perhaps the biggest surprise from the consortium's findings, which were from the results of 300 *years* of computing time, is that a large proportion of dark DNA consists of gene "switches." Over four million gene switches in the noncoding DNA constitute an almost inconceivably intricate information wiring system, one that turns genes on and off and provides a mechanism to rewrite DNA's coded protein structure. (Kolata 2012)

That intricate information wiring system reminds me of the A. C. Gilbert Erector set that I was fascinated with as a child. A parent's nightmare, this kit contained hundreds of parts, including nuts and bolts, various metal beams with regular holes for assembly, pulleys, wheels, gears, and a small electric motor. What distinguishes construction sets like the Erector and today's LEGO blocks is the user's ability to build a model and then take it apart and build something completely different with the same parts, endlessly.

In an analogous biology construction set, genes are the physical building parts, and noncoding DNA is the "instruction sheet" on how to assemble specific models (i.e., animals and plants) from an assortment of the same parts. Like the Erector set, gene-derived protein parts can be assembled, disassembled, and reassembled into a variety of different organisms. The body plans for each organism, encoded in the dark DNA, are directly connected to the dynamic environment via epigenetic mechanisms that interpret, translate, and control the activity of the protein-coding genes.

The results of the ENCODE project are radically changing the research paradigm associated with diseases like cancer. Before new insights about the role of dark DNA surfaced, scientists studying the genetics of disease only sought to identify mutations in the genome's protein-coding genes. The ENCODE assessments now reveal that as many or more disease-associated mutations are present in the dark matter, the noncoding DNA. (Hall 2012) When researchers combine data from the Human Genome Project and the ENCODE project, they are able to identify noncoding DNA stretches called "ultrasensitive" regions. These regions of the so-called junk DNA show almost the same levels of mutations as those in protein-coding genes. When researchers read the genomes of ninety cancer patients, including those with breast cancer, prostate cancer, and brain tumors, they discovered nearly a hundred mutations in noncoding, ultrasensitive regions that were directly involved with the cancer. These first dark DNA investigations focused on cancer research, but the influence of noncoding DNA on other disease is now being studied as well. (Khurana, et al, 2013)

While research over the last decade has provided amazing insights into the structure and function of the genome's noncoding elements, i.e., its "junk" DNA, it has also provided amazing insights into how the 2 percent of the genome that encodes protein impacts health and disease. These studies focus on stretches of DNA called *telomeres* that extend from the ends of protein-coding genes. Though telomere DNA is noncoding—it does not contribute to the structure of the protein's gene blueprint—it provides two vital functions.

First, telomere extensions physically prevent the DNA double helix from unwinding. This is an important function because structurally unstable, "frayed" DNA compromises the coded information needed to assemble a functional protein. Functionally, telomeres resemble the plastic tips at the ends of a shoelace, known in the trade as *aglets*. When an aglet fails, the woven threads at the tip of the shoelace unwind and become frayed, which impairs its function. How many times has your patience been tried guiding a fat, frayed shoelace into a teeny, tiny eyelet?

Second, telomeres provide the physical platform required for DNA replication. A cell must duplicate its DNA before it divides to ensure that each daughter cell receives a complete genome. In this process, an enzyme (DNA helicase) unzips the double helix while a large protein complex, DNA polymerase, attaches onto the free end of the DNA strand. The polymerase enzyme travels like a train on a track down the length of the DNA. As it does so, it assembles a complementary strand of DNA in its wake. However, when the polymerase "train" reaches the end of the DNA strand, it runs into a technical problem . . . the enzyme cannot duplicate the length of DNA on which it sits (see illustration). Consequently, every time a DNA strand is duplicated, it is shorter than the previous copy because the "terminal" (i.e., the last piece of DNA beneath the polymerase enzyme) is not duplicated.

Telomeres prevent a loss of protein-encoding information during gene replication by providing a noncoding stretch of DNA whose loss will not affect the protein's blueprint. This extra length of DNA allows the polymerase "train" to lose a piece of DNA without compromising the region containing the protein code. The length of the telomere extension determines how many times DNA can be copied before polymerase clipping cuts into the gene's protein code. When frequent cell divisions deplete telomere extensions, subsequent copies of the DNA produce dysfunctional proteins.

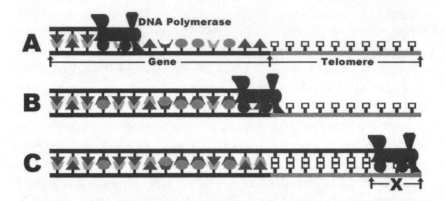

Replication of DNA. Before DNA is copied, the double helix is split into two separate helical strands. In figure A, DNA polymerase, an enzyme that copies the DNA, is represented by the train engine. The polymerase enzyme travels down the length of a single strand of DNA. The gene-coding section of the DNA strand, represented by the black "train track," has a sequence of bases that code for the protein. The telomere section of the DNA, represented by the gray portion of the "train track," has a sequence of noncoding DNA (white "boxes"). As the polymerase moves down the DNA it assembles a complementary DNA strand in its wake. In figure B, the length of the new complementary DNA strand is longer as the polymerase copies more of the original strand. In figure C, the polymerase reaches the end of the DNA strand ("track"). The new complementary DNA molecule is complete. However, it is shorter than the original DNA template because the polymerase enzyme cannot copy the section of DNA on which it sits (X). Each time the DNA is copied, the new DNA strand is shorter than the previous version. After a number of cell divisions, the telomere extension is eliminated and the polymerase begins to clip off pieces of DNA that contain the protein's code. Proteins synthesized from a shortened DNA code are defective and can cause the cell to become dysfunctional.

As defective proteins accumulate, the cells malfunction and ultimately die, but that doesn't have to happen quickly! In the 1960s, Leonard Hayflick calculated from his observations on cultured cells that they could safely divide for approximately fifty generations before their telomeres are lost and subsequent DNA replication produces defective proteins that compromise the cell's health and its ability to further divide. His insights about telomeres led to the belief that humans have a limited lifespan determined by how many times stem cells divide when replacing the billions of cells that die every day. (Hayflick 1965)

Before you get depress[...]
stem cells continue to divide[...]
have identified a special enz[...]
to extend telomere length. T[...]
alent of the "fountain of yout[...]
increase the vitality and repro[...]
ity enhances health and exter[...]

But there is a catch! Life[...]
telomerase activity. For examp[...]
periences, childhood abuse (b[...]
lence, post-traumatic stress di[...]
and lack of love all inhibit telom[...]
to the onset of disease and a [...]rtened life span. In contrast, exercise, good nutrition, a positive outlook on life, living in happiness and gratitude, being in service, and experiencing love, especially self-love, all enhance telomerase activity and promote a long and healthy life. (Blackburn and Epel 2012, Stetka 2014) In fact, a recent Canadian study found that breast cancer patients who were involved in a support group and mindfulness meditation preserved telomere length while the telomeres of a control group without those interventions became shorter. (Carlson, et al, 2014)

As I'll talk about in more depth in later chapters, the primary influence controlling telomerase activity is the mind, which is influenced by the programming we acquired before age seven. And, as I'll discuss, YES . . . we can consciously empower ourselves by actively enhancing our own telomerase. And, YES, because I can't repeat this enough, factoring in all the wonderful research that has been done in the last decade: it's the environment, stupid!

CHAPTER 3

THE MAGICAL MEMBRANE

Now that we've looked at the protein assembly machinery of the cell, debunked the notion that the nucleus is the brain of the cellular operation, and recognized the crucial role the environment plays in the operation of the cell, we're on to the good stuff—the stuff that can make sense of your life and give you insight into ways of changing it.

This chapter puts forth my nominee for the true brain that controls cellular life—the membrane. I believe that when you understand how the chemical and physical structure of the cell's membrane works, you'll start calling it, as I do, the magical membrane. Or alternatively, capitalizing on the fact that part of the word is a homophone for brain, I refer to it in my lectures as the magical mem-Brain. And when you couple your understanding of the magical membrane with an understanding of the exciting world of quantum physics that I'll present in the next chapter, you will also understand how wrong the tabloids were in 1953. The secret of life does not lie in the famed double helix. Insight into the secret of life lies in understanding the elegantly simple biological mechanisms of the magical membrane—the mechanisms by which your body translates environmental signals into behavior.

When I started studying cell biology in the 1960s, the idea that the membrane was the cell's brain would have been considered laughable. And I have to concede that the membrane in those days was a sorry-looking Mensa candidate. The membrane seemed to be just a simple, semi-permeable, three-layered skin that held the contents of the cytoplasm together. Think plastic wrap with holes.

One reason scientists misjudged the membrane is that it is so thin. Membranes are only seven millionths of a millimeter thick. In fact, they are so thin that they can only be seen with an electron microscope, which was developed after the Second World War. So it wasn't until the 1950s that biologists could even confirm that cell membranes exist. Up until that time, many biologists thought the cell's cytoplasm held together because it had a Jell-O-like consistency. With the aid of microscopes, biologists learned that *all* living cells have membranes and that all cell membranes share the same basic, three-layered structure. However, the simplicity of that structure belies its functional complexity.

Cell biologists gained insight into the amazing abilities of the cell membrane by studying the most primitive organisms on this planet, the prokaryotes. Prokaryotes, which include bacteria and other microbes, consist only of a cell membrane that envelops a droplet of soupy cytoplasm. Though prokaryotes represent life in its most primitive form, they have purpose. A bacterium does not bounce around in its world like a ball in a pinball machine. A bacterium carries out the basic physiologic processes of life like more complicated cells. A bacterium eats, digests, breathes, excretes waste matter, and even exhibits "neurological" processing. They can sense where there is food and propel themselves to that spot. Similarly, they can recognize toxins and predators and purposely employ escape maneuvers to save their lives. In other words, pro-karyotes display intelligence!

So what structure in the prokaryotic cell provides its "intel-ligence"? The prokaryotes' cytoplasm has no evident organelles, such as the nucleus and mitochondria, that are found in more advanced, eukaryotic cells. The most likely candidate for the prokaryote's brain is its cell membrane, the only organelle found in every living cell.

Bread, Butter, Olives, and Pimentos

As I came to the realization that membranes were characteristic of all intelligent life, I focused my attention on understanding
their structure and function. I came up with a gastronomic treat
(just kidding) to illustrate the basic structure of the membrane.
The treat consists of a bread and butter sandwich. To further refine
the analogy, I added olives. Actually my instructive sandwich fea-
tures two kinds of olives, some stuffed with pimentos, the others
pimento-free. Gourmands, don't groan. When I've left this sandwich out of my lectures, repeat members of the audience have asked
me where it went!

Here's an easy experiment to show you how the "sandwich"
membrane works. Make a bread-and-butter sandwich (at the
moment free of olives). This sandwich represents a section of the
cell membrane.

Now pour a teaspoon of colored dye on top of the sandwich.

As illustrated below, the dye seeps through the bread but stops when it gets to the butter because the oily substance in the middle of the sandwich provides an effective barrier.

Now let's make a bread and butter sandwich with stuffed and unstuffed olives.

Now when we add the dye to the bread and slice the sandwich, we see a different result. When the dye hits a pimento-stuffed olive, it stops as surely as it stopped when it hit butter. But when the dye reaches an olive without a pimento, the pitted olive provides a channel through which the dye can flow freely across the middle of the sandwich, then through the bread to the plate.

The plate in this analogy represents the cell's cytoplasm. By passing through the pimento-free olive, the dye penetrates the buttery layer to reach the other side of the "membrane" sandwich. The dye has successfully navigated the formidable, fatty membrane barrier!

It is important for the cell to allow molecules to break through the barrier because in my sandwich analogy, the dye represents life-sustaining food. If the membrane were simply a bread-and-butter sandwich, it would provide a fortress-like barrier that keeps out the cacophony of innumerable molecular and radiant energy signals that make up a cell's environment. But the cell would die if the membrane were such a fortress because it would get no nutrients. When you add the pimento-free olives, which allow information and food into the cell, the membrane becomes a vital and ingenious mechanism enabling selected nutrients to penetrate the interior of the cell, just as the teaspoonful of dye made its way to the plate.

In real-life cellular biology, the bread-and-butter portion of the sandwich represents the membrane's phospholipids, one of the two major chemical components of the membrane. (The other major chemical components are the "olive" proteins, which we'll get to below.) I call phospholipids "schizophrenic" because they are composed of both polar and nonpolar molecules.

The fact that phospholipids contain both polar and nonpolar molecules may not sound like a recipe for schizophrenia to you, but I assure you it is. All the molecules in our universe can be divided into nonpolar and polar categories based on the type of chemical bonds that hold their atoms together. The bonds among polar molecules have positive and/or negative charges, hence their polarity. These molecules' positive and negative charges cause them to behave like magnets, attracting or repelling other charged molecules.

Polar molecules include water and things that dissolve in water. Nonpolar molecules include oil and substances that dissolve in oil; there are no positive or negative charges among their atoms. Remember the adage "water and oil don't mix"? Neither do oily nonpolar and watery polar molecules. To visualize the lack of inter-action between polar and nonpolar molecules, think of your bottle of Italian salad dressing. You do your best to get vinegar and oil to bond by shaking the bottle, but when you set the bottle down, they separate. That's because molecules, like people, prefer environments that offer them stability. For their stability, polar (vinegar) molecules seek out watery polar environments and nonpolar (olive oil) molecules seek out nonpolar environments. Phospholipid molecules, comprised of both polar and nonpolar lipid regions, have a difficult time in seeking stability. The polarized phosphate portion of the molecule is motivated to seek water, while its nonpolar lipid portion abhors water and seeks stability by dissolving in oil.

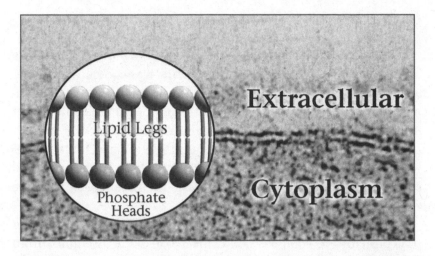

Electron micrograph showing the cell membrane at the surface of a human cell. The dark-light-dark layering of the cell membrane is due to the ordering of the barrier's phospholipid molecules (inset). The lighter center of the membrane, the equivalent of the butter in our sandwich, represents the hydrophobic zone formed by the nonpolar legs of the phospholipids. The dark layers above and below the central lipid zone, the equivalent of the bread slices, represent the molecule's water-loving phosphate heads.

Getting back to our sandwich, the membrane's phospholipids are shaped like lollipops with an extra stick (see illustration above). The round part of the lollipop has polar charges among its atoms; it corresponds to the bread of our sandwich. The molecule's two stick-like portions are nonpolar; they correspond to the butter part of our sandwich. Because the "butter" portion of the membrane is nonpolar, it does not let positively or negatively charged atoms or molecules pass through it. In effect, this lipid core is an electrical insulator, a terrific trait for a membrane designed to keep the cell from being overwhelmed by every molecule in its environment.

But the cell could not survive if the membrane were the equivalent of a simple bread-and-butter sandwich. Most of the cell's nutrients consist of charged polar molecules that would not be able to get past the formidable nonpolar lipid barrier. Neither could the cell excrete its polarized waste products.

Integral Membrane Proteins

The olives in our sandwich are the truly ingenious part of the membrane. These proteins allow nutrients, waste materials, as well as other forms of "information" to be transported across the membrane. The protein "olives" allow not just any old molecules to get into the cell but only those molecules necessary for the smooth functioning of the cytoplasm. In my sandwich, the olives represent Integral Membrane Proteins (IMPs). These proteins embed themselves into the "butter" layer of the membrane, just as I have embedded olives in the illustration.

How do IMPs embed themselves into the butter? Remember that proteins are composed of a linear backbone assembled from linked amino acids. Of the twenty different amino acids, some are water-loving (hydrophylic), polar molecules and some are water-fearing (hydrophobic), nonpolar molecules. When a region of the protein's backbone is made up of linked, hydrophobic amino acids, this segment of the protein seeks stability by finding an oil-loving environment like the membrane's lipid core (see arrow below). That's how hydrophobic parts of the protein integrate themselves into the middle layer of the membrane. Because some regions of a protein's backbone are made up of polar amino acids and other regions are nonpolar, the protein strand will weave itself in and out of the bread-and-butter sandwich.

There are lots of IMPs with lots of different names, but they can be subdivided into two functional classes: *receptor proteins* and *effector proteins.* Receptor IMPs are the cell's sense organs, the equivalent of our eyes, ears, nose, taste buds, etc. Receptors function as molecular "nano-antennas" tuned to respond to specific environmental signals. Some receptors extend inward from the membrane surface to monitor the internal milieu of the cell. Other receptor proteins extend from the cell's outer surface, monitoring external signals.

Like other proteins, which we discussed earlier, receptors have an inactive and an active shape and shift back and forth between those conformations as their electrical charges are altered. When a receptor protein binds with an environmental signal, the resulting alteration in the protein's electrical charges causes the backbone to change shape and the protein adopts an "active" conformation. Cells possess a uniquely "tuned" receptor protein for every environmental signal that needs to be read.

Some receptors respond to physical signals. One example is an estrogen receptor, which is specially designed to complement the shape and charge distribution of an estrogen molecule. When estrogen is in its receptor's neighborhood, the estrogen receptor locks on to it, as surely as a magnet picks up paper clips. Once the estrogen receptor and the estrogen molecule bind in a perfect "lock and key" fit, the receptor's electromagnetic charge changes and the protein shifts into its active conformation. Similarly, histamine receptors complement the shape of histamine molecules, and insulin receptors complement the shape of insulin molecules.

Receptor "antennas" can also read vibrational energy fields such as light, sound, and radio frequencies. The antennas on these "energy" receptors vibrate like tuning forks. If an energy vibration in the environment resonates with a receptor's antenna, it will alter the protein's charge, causing the receptor to change shape. (Tsong 1989) I'll cover this more completely in the next chapter, but I'd like to point out now that because receptors can read energy fields, the notion that only physical molecules can impact cell physiology is outmoded. Biological behavior can be controlled by invisible forces, including thought, as well as it can be controlled by physical molecules like penicillin, a fact that provides the scientific underpinning for pharmaceutical-free energy medicine.

Receptor proteins are remarkable, but on their own they do not impact the behavior of the cell. While the receptor provides an awareness of environmental signals, the cell still has to engage in an appropriate, life-sustaining response; that is the venue of the effector proteins. Taken together, the receptor-effector proteins are a stimulus-response mechanism comparable to the reflex action that doctors typically test during physical examinations. When a doctor taps your knee with a mallet, a sensory nerve picks up the signal. That sensory nerve immediately passes on that information to a motor nerve that causes the leg to kick. The membrane's receptors are the equivalent of sensory nerves, and the effector proteins are the equivalent of action-generating motor nerves. Together, the receptor-effector complex acts as a switch, translating environmental signals into cellular behavior.

It is only in the last twenty years that scientists have realized the importance of the membrane's IMPs. They are in fact so important that studying the way IMPs work has become a field of its own called "signal transduction." Signal transduction scientists are busily classifying hundreds of complex information pathways that lie between the membrane's reception of environmental signals and the activation of the cell's behavior proteins. The study of signal transduction is catapulting the membrane to center stage, just as the field of epigenetics is highlighting the role of the chromosome's proteins.

There are different kinds of behavior-controlling effector proteins because there are lots of jobs that need to be done for the smooth functioning of the cell. Transport proteins, for example, include an extensive family of channel proteins that shuttle molecules and information from one side of the membrane barrier to the other. Which brings us back to the pimentos in our bread, butter, and olive sandwich. Many channel proteins are shaped like a tightly wound sphere, resembling the pimento-stuffed olives in our pictures. (See illustration page 61.) When the electrical charge on the protein is altered, the protein changes shape, a change that creates an open channel running through the protein's core. Channel proteins are actually two olives in one, depending on their electrical charge. In the active mode, their structure resembles a pimento-free olive, with an open gate. In their inactive mode the proteins' shape resembles a pimento-stuffed olive that stays closed to the world outside the cell.

The activity of one specific channel type, sodium-potassium ATPase, merits special attention. Every cell has thousands of these channels built into the membrane. Collectively, their activity uses almost half of your body's energy every day. This channel opens and closes so frequently that it resembles a revolving door in a department store on the day of a big sale. Every time this channel revolves, it shuttles three positive-charged sodium atoms out of the cytoplasm and simultaneously admits two positive-charged potassium atoms into the cytoplasm from the environment.

Sodium-potassium ATPase not only uses up a lot of energy, it also creates energy as surely as store-bought batteries provide energy for flashlights (at least until you forget to change them before the big storm). Actually, the energy-producing activity of sodium-potassium ATPase is a lot better than the batteries your kids wear out because it turns the cell into a constantly recharging biological battery.

Here's how sodium-potassium ATPase manages that trick. Every revolution of sodium-potassium ATPase throws more positive charges out than it lets in to the cell, and there are thousands of these proteins in each cell membrane. As these proteins go through hundreds of revolution cycles per second, the inside of the cell becomes negatively charged while the outside of the cell becomes positively charged. The negative charge below the membrane is referred to as the *membrane potential*. Of course the lipid, i.e., the butter portion of the membrane, does not let charged atoms cross the barrier, so the internal charge stays negative. The positive charge outside the cell and the negative charge inside make the cell essentially a self-charging battery whose energy is used to empower biological processes.

Another variety of effector proteins, cytoskeletal proteins, regulates the shape and motility of cells. A third variety, called enzymes, breaks down or synthesizes molecules, which is why enzymes are sold in your local health food store as a digestive aid. When activated, all forms of effector proteins, including channels, cytoskeletal, and enzyme proteins or their by-products, can also serve as signals that activate genes. These IMPs or their byproducts provide signals that control the binding of the chromosome's regulatory proteins that form a "sleeve" around the DNA. In contrast to conventional wisdom, genes do not control their own activity. Instead it is the membrane's effector proteins, operating in response to environmental signals picked up by the membrane's receptors that *control* the "reading" of genes so that worn-out proteins can be replaced or new proteins can be created.

How the Brain Works

Once I understood how IMPs worked, I had to conclude that *the cell's operations are primarily molded by its interaction with the environment, not by its genetic code.* There is no doubt that the DNA blueprints stored in the nucleus are remarkable molecules, which have been accumulated over three billion years of evolution. But as remarkable as these DNA blueprints are, they do not "control" the operations of the cell. Logically, genes cannot preprogram a cell or organism's life because cell survival depends on the ability to dynamically adjust to an ever-changing environment.

The membrane's function of interacting "intelligently" with the environment to produce behavior makes it the true brain of the cell. Let's put the membrane to the same "brain" test to which we put the nucleus. When you destroy its membrane, the cell dies just as you would if your brain were removed. Even if you leave the membrane intact, destroying only its receptor proteins, which can easily be done with digestive enzymes in the lab, the cell becomes "brain-dead." It is comatose because it no longer receives the environmental signals necessary for the operation of the cell. The cell also becomes comatose when the membrane's receptor proteins are left intact and its effector proteins are immobilized.

To exhibit "intelligent" behavior, cells need a functioning membrane with both receptor (awareness) and effector (action) proteins. These protein complexes are the fundamental units of cellular intelligence. Technically they may be referred to as units of "perception." The definition of perception is "awareness of the elements of environment through physical sensation." The first part of the definition describes the function of receptor IMPs. The second part of the definition, the creation of a "physical sensation," sums up the role of the effector proteins.

By examining these basic units of perception, we have engaged in an ultimate reductionist exercise, taking the cell down to its fundamental nuts and bolts. In this regard it is important to note that at any given time there are up to hundreds of thousands of such switches in a cell membrane. Consequently, the behavior of a cell cannot be determined by examining any individual switch. The behavior of a cell can only be understood by considering the activities of *all* the switches at any given time. That is a holistic—not reductionist—approach, which I'll elaborate on in the next chapter.

At the cellular level, the story of evolution is largely the story of maximizing the number of basic units of "intelligence," the membrane's receptor-effector proteins. Cells became smarter by utilizing their outer membrane surface more efficiently and by expanding the surface area of their membranes so that more IMPs could be packed in. In primitive prokaryote organisms, the cell membrane's IMPs carry out all of its fundamental physiologic functions including digestion, respiration, and excretion. Later in evolution, portions of the surface membrane that carry out these physiologic functions go inside the cell, forming the membranous organelles that are characteristic of eukaryotic cytoplasm. That leaves more membrane surface area available to increase the number of perception IMPs. In addition, the eukaryote is thousands of times bigger than the prokaryote resulting in a tremendous increase in membrane surface area, i.e., a whole lot more room for IMPs. The end result is more awareness, which translates to greater survivability.

Through evolution, the cell membrane's surface expanded, but there was a physical limit to that expansion. There was a point at which the thin cell membrane was not strong enough to contain a larger mass of cytoplasm. Think what happens when you fill a balloon with water. As long as the balloon is not overfilled, it is strong and can be passed around. However, if you exceed the balloon's water capacity, the balloon ruptures easily, spilling its contents, just as a membrane with too much cytoplasm would inevitably rupture. When the cell membrane reached that critical size, the evolution of the individual cell reached its limit. That's why for the first three billion years of evolution, single cells were the only organisms on this planet. That situation changed only when cells came up with another way to increase awareness. In order to get smarter, cells started banding together with other cells to form multicellular communities through which they could share their awareness, as I explained in Chapter 1.

To review, the functions required for a single cell to stay alive are the same functions required by a community of cells to stay alive. But cells started to specialize when they formed multicellular organisms. In multicellular communities, there is a division of labor. That division of labor is evident in the tissues and organs that carry out specialized functions. For example, in the single cell, respiration is carried out by the mitochondria. In a multicellular organism, the mitochrondrial equivalent for respiration are the billions of specialized cells that form the lungs. Here's another example: In the single cell, movement is created by the interaction of cytoplasmic proteins called actin and myosin. In a multicellular organism, communities of specialized muscle cells handle the job of generating motility, each endowed with massive quantities of actin and myosin proteins.

I repeat this information from the first chapter because I want to emphasize that while it is the job of the membrane in a single cell to be aware of the environment and set in motion an appropriate response to that environment, in our bodies those functions have been taken over by a specialized group of cells we call the nervous system. It is not a coincidence that the human nervous system is derived from the embryonic skin, the human counterpart of a cell's membrane.

Though we've come a long way from unicellular organisms, I believe, as I've mentioned before, that studying single cells is an instructive way of studying complicated multicellular organisms. Even the most complex human organ, the brain, will reveal its secrets more readily when we know as much as we can about the membrane, the cell's equivalent of a brain.

The Secret of Life

As you've learned in this chapter, scientists have recently made great progress toward unraveling the complexity of the simple-looking membrane. But even twenty-five years ago, the rough outlines of the membrane's functions were known. In fact, it was 1985 when I first realized how understanding the workings of the membrane could be life changing. My eureka moment resembled the dynamics of super-saturated solutions in chemistry. These solutions, which look like plain water, are fully saturated with a dissolved substance. They are so saturated that adding just one more drop of the substance causes a dramatic reaction in which all of the dissolved materials instantly coalesce into a giant crystal.

In 1985, I was living in a rented house on the spice-drenched Caribbean island of Grenada teaching at yet another "off-shore" medical school. It was 2 A.M., and I was up revisiting years of notes on the biology, chemistry, and physics of the cell membrane. At the time I was reviewing the mechanics of the membrane, trying to get a grasp of how it worked as an information processing system. That is when I experienced a moment of insight that transformed me, not into a crystal, but into a membrane-centered biologist who no longer had any excuses for messing up his life.

At that early morning hour, I was redefining my understanding of the structural organization of the membrane, starting first with the lollipop-like phospholipid molecules and noting that they are arranged in the membrane like regimented soldiers on parade in perfect alignment. By definition, a structure whose molecules are arranged in regular, repeated pattern is a crystal. There are two fundamental types of crystals. The crystals that most people are familiar with are hard and resilient minerals like diamonds, rubies, and even salt. The second kind of crystal has a more fluid structure even though its molecules maintain an organized pattern. Familiar examples of *liquid crystals* include digital watch faces and laptop computer screens.

To better understand the nature of a liquid crystal, let's go back to those soldiers on parade. When the marching soldiers turn a corner, they maintain their regimented structure, even though they're moving individually. They're behaving like a flowing liquid, yet they do not lose their crystalline organization. The phospholipid molecules of the membrane behave in a similar fashion. Their fluid crystalline organization allows the membrane to dynamically alter its shape while maintaining its integrity, a necessary property for a supple membrane barrier. So in defining this character of the membrane I wrote: "The membrane is a liquid crystal."

Then I started thinking about the fact that a membrane with just phospholipids would be simply a bread-and-butter sandwich without the olives. In the experiment described earlier, the colored dye could not get through the lipid butter layer. That bread-and-butter sandwich is a nonconductor. However, when you include the IMP "olives," you realize that the membrane conducts some things across while keeping other things out. So I continued writing my description of the membrane by adding: "The membrane is a *semiconductor.*"

Lastly, I wanted to include in my description the two most common kinds of IMPs. These are the receptors and a class of effectors called channels because they provide the all-important means for the cell to let in nutrients and let out waste matter. I was about to write that the membrane contains "receptors and channels" when I realized that a synonym for receptor is the word "gate." So instead I completed my description by writing: "The membrane contains *gates* and *channels.*"

I sat back and reviewed my new description of the membrane: *"The membrane is a liquid crystal semiconductor with gates and channels."* What hit me right away was the fact that I had recently heard or read the very same phrase, though at the moment, I didn't know where I had come across it. One thing was for sure; it was not in the context of biological science.

As I leaned back in my chair, my attention was drawn to the corner of my desk where my new, smiley-face Macintosh, my first computer, was parked. Lying beside the computer was a copy of a bright red book called *Understanding Your Microprocessor.* I had just bought this nontechnical paperback guide to how computers work from a Radio Shack outlet. I grabbed the book and found in the introduction a definition of a computer chip that read: "A chip is a crystal semiconductor with gates and channels."

For the first second or two I was struck by the fact that the chip and cell membrane shared the same technical definition. I spent several more intense seconds comparing and contrasting biomembranes with silicon semiconductors. I was momentarily stunned when I realized that the identical nature of their definitions was not a coincidence. The cell membrane was indeed a structural and functional equivalent (homologue) of a silicon chip!

Twelve years later, in 1997, an Australian research consortium headed by B. A. Cornell published an article in *Nature* that confirmed my hypothesis that the cell membrane is a homologue of a computer chip. (Cornell, et al, 1997) The researchers isolated a cell membrane and attached a piece of gold foil under it. They then flooded the space between the gold foil and the attached membrane with a special electrolyte solution. When the membrane's receptors were stimulated by a complementary signal, the channels opened and allowed the electrolyte solution to flow across the membrane. The gold foil served as a transducer, an electrical pickup device, which converted the electrical activity of the channel into a digital readout on a screen. This device, created for the study, demonstrates that the cell membrane not only looks like a chip but also functions like one. Cornell and associates successfully turned a biological cell membrane into a digital-readout computer chip.

So what's the big deal, you ask? The fact that the cell membrane and a computer chip are homologues means that it is both appropriate and instructive to better fathom the workings of the cell by comparing it to a personal computer. The first big-deal insight that comes from such an exercise is that computers and cells are *programmable*. The second corollary insight is that the programmer lies *outside* the computer/cell. Biological behavior and gene activity are dynamically linked to information from the environment, which is downloaded into the cell.

The point: a cell is a "programmable chip" whose behavior and genetic activity are primarily controlled by environmental signals, not genes.

I had been trained as a nucleus-centered biologist as surely as Copernicus had been trained as an Earth-centered astronomer, so it was with a jolt that I realized that the gene-containing nucleus does not program the cell. Environmental data is entered into the cell/computer via the membrane's receptors, which represent the cell's "keyboard." Receptors trigger the membrane's effector proteins, which act as the cell/computer's "Central Processing Unit" (CPU).

The function of the computer's CPU is to convert incoming data into the binary code language used by the computer's operating system. The receptor-effector protein complexes represent a functional complement of a computer's CPU processor. Incoming environmental information is passed from the receptor to the effector protein, which in turn converts the incoming signal into the behavioral language of biology.

I realized in those early morning hours that even though biological thought at that time was still preoccupied with genetic determinism, leading-edge cell research, which continues to unfold the mystery of the Magical Membrane in ever more complex detail, tells a far different story.

At that moment of transformation, I was frustrated because there was no one with whom I could share my excitement. I was alone out in the country. My house didn't have a telephone. Because I was teaching at a medical school, I realized that there would undoubtedly be some students studying in the library. I hastily threw some clothes on and raced off to the school to tell someone, anyone, of this exciting new insight.

Running into the library, out of breath, wild-eyed with my hair flying in all directions, I was the epitome of the absent-minded professor. I spotted one of my first-year medical students and ran up to him proclaiming, "You have to hear this! This is great shit!" I remember in the back of my mind how he pulled away from me, almost in fear of this raving, mad scientist who wildly broke the silence of the sleepy library. I immediately began to spew forth my new understanding of the cell, using the complex, polysyllabic jargon of a conventional cell biologist. When I finished my explanation and was silent, I was waiting to hear his congratulations or at least a "bravo," but nothing was forthcoming. He was now wide-eyed himself. All he could say was, "Are you okay, Dr. Lipton?"

I was crushed. The student had not understood a word I had said. In hindsight, I realized that as a first-semester medical student, he did not have enough scientific background or vocabulary to make any sense out of my apparent rantings. However, the wind was knocked out of my sails. I held the key to the secret of life, and there was no one who could understand me! I confess I didn't have much better luck with most of my colleagues who had been schooled in polysyllabic jargon. So much for the Magical Membrane.

Over the years I gradually honed my presentation about the Magical Membrane and continued to refine it so that first-year medical students and lay people can understand it. I've also continued to update it with the latest research. In so doing, I've found much more receptive audiences among a wider range of people. I have also found audiences receptive to the spiritual implications of my eureka moment. Shifting to membrane-centered biology was exciting for me, but it wouldn't have been enough to send me screaming to the library. That Caribbean moment not only transformed me into a membrane-centered biologist, it also transformed me from an agnostic scientist into a card-carrying mystic who believes that eternal life transcends the body.

I'll get to the spiritual part of the story in the Epilogue. For the moment, let me reiterate the lessons of the Magical Membrane, which put the control of our lives not in the genetic roll of the dice at conception but in our own hands. We are the drivers of our own biology, just as I am the driver of this word processing program. We have the ability to edit the data we enter into our biocomputers, just as surely as I can choose the words I type. When we understand how IMPs control biology, we become masters of our fate, not victims of our genes.

✳ ✳ ✳

I can't say that mainstream scientists have taken on my spelling of mem-Brain and neither that they have been touting my message that Integral Membrane Proteins (IMPs) make us masters of our fate. But

I can say that there is a wealth of new and fully supportive research about how the membrane interacts with the environment to shape biology.

For example, research on the cell's membrane potential, which I talked about above, has opened up a new way of thinking for developmental biologists who had previously only focused on the role of signal *molecules* (hormones, neurotransmitters, or other chemical agents) in controlling development and creating body parts. In 2011, biologist Michael Levin's team at the Tufts Center for Regenerative and Developmental Biology altered the bioelectrical voltage of the membranes of tadpole cells. Amazingly, simply altering the membrane potential in cells from the backs and tails of tadpoles resulted in fully developed eyes growing in the backs and tails, far from where eyes normally form. (Pai, et al, 2011) Sounds like impressive, magical mem-Brains to me!

The key to the success of the study was the team's finding that during the embryonic development of a tadpole, the membrane potential in cells destined to form an eye drops precipitously from about -70 millivolts to about -20 millivolts. In their lab, Levin's group induced the same drop to -20 millivolts by inserting voltage-regulating calcium ion channel proteins into the membranes of the tadpoles' back and tail cells, triggering a signal that initiated the growth of a complete eye. This research is exciting because it opens the possibility of repairing birth defects and regenerating damaged human organs. It also underscores the fact that the membrane controls cell behavior using nonchemical, "electric" (more next chapter!) environmental signals: "Aside from the regenerative medicine applications of this new technique for eyes, this is the first step to cracking the bioelectric code," said Levin. (Yuhas 2013)

Membrane research is also helping rehabilitate the reputation of cholesterol, which has long been vilified as the culprit in the modern-day scourges of cardiac disease, heart attacks, and stroke. In a case of guilt by association, high levels of cholesterol are present in 35 percent of patients with cardiovascular disease, and at sites of vascular damage, the endothelial cells, the inner lining of blood vessels, are loaded with cholesterol droplets. But I want to offer a more nuanced view of cholesterol, which is often lost in the rush to demonize it.

Cholesterol is a lipid molecule that plays a vital role in our day-to-day survival. It is, for example, the precursor for the synthesis of important steroid molecules, including the bile salts used in digestion, regulatory steroid hormones such as estrogen and cortisol, and vitamin D.

More relevant to this chapter, cholesterol is an essential component of the membrane whose function is required for the survival of our 50 trillion cells, which is another way of saying *our* survival. Cholesterol helps the membrane maintain a very important balancing act: it must be rigid enough to physically resist the strain placed on it by the cytoplasm it encloses, yet supple enough to accommodate the flexibility required for the movement of cells.

Membrane fluidity is also of great importance in controlling the cell's "brain" function because it impacts the membrane's ability to read and respond to environmental information. To function normally, IMPs, in the form of receptor and effector molecules, must be able to engage one another by freely circulating within the membrane's inner, oil-loving, hydrophobic core. It is the viscosity of the membrane's lipid core that controls the ability of these proteins to circulate freely. A membrane made up only of phospholipid molecules would be quite fluid, so it would enhance the mobility of IMPs, but it would not be rigid enough to hold up under the pressure generated by the enclosed cytoplasm.

Cholesterol is a more rigid molecule than phospholipid. So when cholesterol is inserted into the membrane, it immobilizes surrounding phospholipid molecules, creating the extra rigidity that strengthens the membrane and impedes the flow of small ions and molecules into the cell. The inserted cholesterol also creates extra space among the phospholipid molecules, space that keeps them from "gelling" so that phospholipid molecules don't turn from an oil-like substance into a butter-like substance. So in addition to stiffening the membrane, cholesterol acts like "antifreeze" that ensures that proteins and lipids are able to move freely. (Holthuis and Menon 2014)

Counterintuitively, the membrane's rigid cholesterol molecules can also act to *restrict* IMP movement. When clusters of cholesterol molecules link up with a class of lipids called sphingolipids, they form structurally rigid "rafts" that restrict the movement of entrapped IMPs. This restriction on IMP movement offers another example of the power of banding together for the greater good. The rafts behave like "corrals" that group clusters of IMPs so they can work together to control specific cellular functions. Cholesterol rafts are the cell's equivalent of short-term memory because the IMPs they contain represent information that engages a variety of cell behaviors. (Korade and Kenworthy 2008)

These vital roles suggest that cholesterol should not be considered a feared evil villain but instead just a foot soldier doing its duty somewhere down the line of command. I was never predisposed to convicting cholesterol for *causing* heart disease because when I was transitioning from the laid-back life of a Caribbean lecturer to the far more frenetic life of a scientist at research juggernaut Stanford, I spent time at what I describe jokingly as a research halfway house at Pennsylvania State University in the lab of Theodore M. Hollis, a gifted scientist I met when he guest-lectured at the island med school.

When I was in his lab, Ted showed me samples of blood from the specialized strain of rats he used to study human atherosclerosis, the hardening and narrowing of arteries that is the leading cause of death in the United States. These animals had so much cholesterol in their systems that their blood was milky white. Despite their apparently toxic level of cholesterol, these rats did *not* form endothelial cell plaques typical of atherosclerotic blood vessels. The secret . . . Ted added an over-the-counter antihistamine drug (the same kind that allergy sufferers turn to regularly) when he introduced the cholesterol. Because the antihistamines could override cholesterol's apparent role in atherosclerotic plaque formation, his work showed that the mere presence of the cholesterol was *not* the driving force behind a blood vessel's malfunction.

Since antihistamines protected the rats, Ted's research obviously suggested an alternative culprit: histamine. (Note: Despite my friend Ted's exciting research on rats, I am not advocating that humans load up on antihistamines! The research is too preliminary for that, and, as you know, I think that all too often, biomedicine rushes to drug solutions without fully understanding their side effects.) Histamine is a stress-related hormone that prepares the body to deal with anticipated injuries and inflammation when the fight-or-flight response is activated by a perceived stressor. Now decades later, the role of histamine in facilitating atherosclerosis has been confirmed. In recent mice studies, the genes for histamine synthesis were experimentally "knocked out." These genetically modified mice, unable to synthesize histamine, resisted the influence of stressors that led to inflammation and atherosclerosis in control mice. And the protective results observed in histamine-free mice were independent of serum cholesterol levels. (Wang, et al, 2011) The results of animal studies point to the role that chronic stress plays in the creation of histamine and in the onset and exacerbation of atherosclerosis and promotion of cardiovascular disease. In direct contrast to the implied role of cholesterol in causing heart disease, cardiovascular pathology may instead primarily result from environmental stressors rather than genetic or biochemical dysfunctions.

Though this research argues against the medical establishment's rush to judgment against cholesterol, that rush was fueled by the interests of the pharmaceutical industry. Of course, that's because the drug companies had come up with another one of their beloved magic bullets, this time in the form of statins. Statins are a class of drugs used to lower levels of cholesterol in the blood by inhibiting a liver enzyme responsible for producing 70 percent of the body's cholesterol. Statin drugs were originally intended for high-risk cardiac patients, but someone, likely in sales, came up with the idea that statins might be good for primary prevention to help those at risk of developing heart disease in the future as well.

The JUPITER Study, frequently referenced in support of statin use, found that during the study period, there were sixty-eight heart attacks in the placebo group and only thirty-one heart attacks in the group that took statins. So according to those numbers, statins produced an astonishing 58 percent reduction in relative risk. The results led the research group to advise that statins were effective for primary prevention of heart attacks. (Ridker 2008) On the surface those stats sound very impressive, but the suggestive conclusion was simply a manipulation of the data. It should be noted that the experimental and control study groups each had 8,901 participants. In real terms, the heart attack risk went from a very low 0.76 percent (68 out of 8,901) in the control group to 0.35 percent (31 out of 8,901) in the statin group. Statistically, the "protective" effect of statins provided for a 0.35 percent reduction over controls, which meant a real risk reduction of less than one half of one percent. The data indicate that for every 300 people taking expensive statin drugs, only one life might be saved. Follow-up studies reveal that the presumed preventive effects of cholesterol-lowering drugs have been considerably exaggerated. As a side note, AstraZeneca, makers of the statin drugs used in the study, was the source of funding for the now discredited JUPITER Study. (Lorgeril, et al, 2010)

The use of statins in the primary prevention of heart disease has fueled statin sales, but it hasn't turned the tide in the war on cardiovascular disease. In fact, as with many wars waged lately, the cost is high and the results negligible. Though statins accounted for $29 billion in U.S. sales in 2013 alone, their war against cholesterol has barely had an impact on cardiovascular diseases. At best, statin drugs lower the actual risk of heart attack by around 0.3 percent, while at the same time producing side effects in 15 percent to 40 percent of those using the drug. Recent independent studies have shown that statin use for primary prevention has minimal or no value in reducing heart attacks and mortality. (Sultan and Hynes 2013)

The statin approach to treating cardiac disease is yet another expensive war with a very poor prognosis. As of yet, we haven't found the "weapons" of mass-cardiac destruction. Instead of continuing to search, perhaps it's time (actually, way past time) to revise the conclusion that cholesterol is culpable for cardiovascular health issues and shift our attention to environmental stressors rather than genetic or biochemical dysfunctions.

The origin of 90 percent of cardiovascular disease is not due to an organic dysfunction in the cell's mechanisms, but rather represents a behavioral response driven by environmental signals in the blood. The brain secretes blood-borne hormones, stress factors, and inflammatory agents in order to coordinate the function of 50 trillion cells to sustain life. This insight returns us full circle to the story of the magical mem-Brain because the cell membrane is the information processor that provides the interface between biology and our brain's perception of the environment. A more complete understanding of cholesterol's vital role in membrane information processing makes it apparent that disturbing cholesterol metabolism with statin drugs is tantamount to throwing yet another monkey wrench into the machine.

Until recently, disease was perceived as a consequence of a break-down in cells' biochemical mechanisms. The vast majority of disease is now recognized to be the result of lifestyle. When biomedicine fully appreciates that the membrane is a truly magical brain, it will shift away from genetic/biochemical dysfunctions and focus on how we can change our perceptions/beliefs (more in later chapters) to prevent cardiac and, in fact, almost all other diseases.

CHAPTER 4

THE NEW PHYSICS:
Planting Both Feet Firmly on Thin Air

When I was an ambitious undergraduate biology major in the 1960s, I knew that to have a prayer of getting into a prestigious graduate school I needed to take a physics course. My college offered a basic introductory course, something like Physics 101, which covered fundamental topics like gravity, electromagnetism, acoustics, pulleys, and incline planes in a way that was easily understood by nonphysics majors. There was also another course called Quantum Physics, but almost all of my peers avoided it like the plague. Quantum physics was shrouded in mystery—we biology majors were convinced that it was a very, very "weird" science. We thought only physics majors, masochists, and outright fools would risk five credits on a course whose premise was: "Now you see it. Now you don't."

In those days the only reason I would have been able to come up with for taking a quantum physics course was that it would have served as a great pickup line at parties. In the days of Sonny and Cher it would have been *très chic* to say, "Hey, babe, I'm into quantum physics. What sign are you?" On the other hand, even that might not be true—I never saw quantum physicists at parties or, in fact, anywhere else. I don't think they got out much.

So I reviewed my transcripts, weighed the options, and took the easy way out by selecting Physics 101. I was intent on becoming a biologist. I had no interest in having my career aspirations depend on some slide-rule-slinging physicist singing the praises of ephemeral bosons and quarks. I and virtually every other biology major either paid little attention to or completely ignored quantum physics as we pursued our studies in the life sciences.

Unsurprisingly, given our attitude, we biology majors didn't know much about physics, the one with all the equations and mathematics. I knew about gravity—heavy things tend to end up at the bottom and lighter things on top. I knew something about light—plant pigments such as chlorophyll and animal visual pigments such as the rhodopsin in the retina absorb some colors of light and are "blind" to others. I even knew a little about temperature—high temperatures inactivate biological molecules by causing them to "melt" and low temperatures freeze and preserve molecules. I am obviously exaggerating to stress the point that biologists traditionally don't know much physics.

My quantum-physics-deprived background explains why, even when I rejected nucleus-based biology and turned to the membrane, I still didn't understand the full implications of that shift. I knew that Integral Membrane Proteins hook up with environmental signals to power the cell. But because I didn't know anything about the quantum universe, I did not fully appreciate the nature of the environmental signals that start the process.

It wasn't until 1982, more than a decade after I had finished graduate school, that I finally learned how much I had missed when I skipped quantum physics in college. I believe that had I been introduced to the quantum world in college, I would have turned into a biology renegade much earlier. But on that day in 1982, I was sitting on the floor of a warehouse in Berkeley, California, 1,500 miles from home, lamenting the fact that I had seriously compromised my scientific career on a failed attempt to produce a rock 'n' roll show. The crew and I were stranded—we had run out of money after six shows. I had no cash and whenever I offered my credit card, the merchant's credit approval machine displayed a skull and crossbones. We were living on coffee and doughnuts while we proceeded through Elisabeth Kübler-Ross's five stages of grieving over the death of our show: denial, anger, bargaining, depression, and, finally, acceptance. (Kübler-Ross 1997) But at that moment of acceptance, the silence in that darkened concrete tomb of a warehouse was broken by the piercing, electronic screech of a telephone. Despite the phone's incessant, obnoxious signal, the crew and I ignored the caller. It wasn't for us—no one knew where we were.

Finally the manager of the warehouse retrieved the call and restored the blessed silence. In the quiet, still air, I heard the manager respond, "Yes, he's here." I looked up at that moment, from the darkest depth of my life, and saw the phone being extended toward me. It was the Caribbean-based medical school that had hired me two years earlier. The president of the school had spent two days tracking my erratic trail from Wisconsin to California so he could ask me if I would be interested in teaching anatomy again.

Would I be interested? Does a bear relieve himself in the woods? "How soon do you want me?" was my reply. He said, "Yesterday." I told him I would love the job but needed an advance on my salary. The school wired the money that same day, and I split the proceeds with my crew. I then flew back to Madison to prepare for an extended stay in the tropics. I bid farewell to my daughters and hastily packed my clothes and a few household items. Within twenty-four hours I was back at O'Hare Airport waiting for Pan Am's Clipper Ship to the Garden of Eden.

By now you're no doubt wondering what my failed rock 'n' roll career has to do with quantum physics—welcome to my unorthodox lecturing style! For the linear-minded, we're officially back to quantum physics, through which I was delighted to learn that scientists cannot understand the mysteries of the universe using only linear thinking.

Listening to the Inner Voice

While I was waiting for the flight, I realized suddenly that I had nothing to read while strapped into a seat for five hours. Moments before the gate was to close, I left the line and ran down the concourse to a bookstore. The task of selecting one book out of hundreds of choices, while simultaneously envisioning the possibility that my plane's doors would close and leave me behind, almost paralyzed me. In a state of confusion, one book jumped out at me—*The Cosmic Code: Quantum Physics as the Language of Nature* by physicist Heinz R. Pagels. (Pagels 1982) I quickly scanned the jacket and found that it was a quantum physics text written for the lay audience. Stubbornly adhering to the quantum physics phobia I had displayed since college, I immediately put the book down and began to search for something lighter.

As the second hand on my mental stopwatch entered into the red zone, I picked up a self-proclaimed best seller and ran to the cashier. While the clerk was preparing to ring up the best seller, I looked up and saw another copy of Pagels's book on the shelf behind the clerk. Midway through the checkout process, with time running out, I finally broke through my aversion to quantum physics and asked the clerk to add a copy of *The Cosmic Code*.

After I boarded the plane, I calmed down from my adrenalized trip to the bookstore, worked on a crossword puzzle, and then finally settled down to read Pagels's book. I found myself burning through its pages, even though I had to continually back up and read sections over again and again. I read through the flight, the three-hour layover in Miami, and an additional five hours en route to my island paradise. Pagels was completely blowing me away!

Before boarding the plane in Chicago, I had no idea that quantum physics was in any way relevant to biology, the science of living organisms. When the plane arrived in Paradise, I was in a state of intellectual shock. I realized that quantum physics *is* relevant to biology and that biologists are committing a glaring, scientific error by ignoring its laws. Physics, after all, is the foundation for all the sciences, yet we biologists almost universally rely on the outmoded, albeit tidier, Newtonian version of how the world works. We stick to the physical world of Newton and ignore the invisible quantum world of Einstein, in which matter is actually made up of energy and there are no absolutes. At the atomic level, matter does not even exist with certainty; it only exists as a *tendency* to exist. All my certitudes about biology and physics were shattered!

In retrospect, it should have been obvious to me and to other biologists that Newtonian physics, as elegant and reassuring as it is to hyper-rational scientists, cannot offer the whole truth about the human body, let alone the universe. Medical science keeps advancing, but living organisms stubbornly refuse to be quantified. Discovery after discovery about the mechanics of chemical signals, including hormones, cytokines (hormones that control the immune system), growth factors, and tumor suppressors cannot explain paranormal phenomena. Spontaneous healings, psychic phenomena, amazing feats of strength and endurance, the ability to walk across hot coals without getting burned, acupuncture's ability to diminish pain by moving "chi" around the body, and many other paranormal phenomena defy Newtonian biology.

Of course, I considered none of that when I was on medical school faculties. My colleagues and I trained our students to disregard the healing claims attributed to acupuncture, chiropractic, massage therapy, prayer, etc. In fact, we went further. We denounced these practices as the rhetoric of charlatans because we were tethered to a belief in old-style, Newtonian physics. The healing modalities I just mentioned are all based on the belief that energy fields are influential in controlling our physiology and our health.

The Illusion of Matter

Once I finally grappled with quantum physics, I realized that when we so cavalierly dismissed those energy-based practices, we were acting as myopically as the chairman of the physics department at Harvard University, who, as described in *The Dancing Wu Li Masters* by Gary Zukav, warned students in 1893 that there was no need for new Ph.D.'s in physics. (Zukav 1979) He boasted that science had established that the universe is a "matter machine" made up of physical, individual atoms that fully obey the laws of Newtonian mechanics. For physicists, the only work left was to refine its measurements.

Three short years later, the notion that the atom was the smallest particle in the universe fell by the wayside with the discovery that the atom itself is made up of even smaller, subatomic elements. Even more earth-shattering than the discovery of those subatomic particles was the revelation that atoms emit various "strange energies" such as x-rays and radioactivity. At the turn of the twentieth century, a new breed of physicist evolved whose mission was to probe the relationship between energy and the structure of matter. Within another ten years, physicists abandoned their belief in a Newtonian, material universe because they had come to realize that the concept of matter is an illusion, for they now recognized that everything in the Universe is made out of energy.

Quantum physicists discovered that physical atoms are made up of vortices of energy that are constantly spinning and vibrating; each atom is like a wobbly spinning top that radiates energy. Because each atom has its own specific energy signature (wobble), assemblies of atoms (molecules) collectively radiate their own identifying energy patterns. So every material structure in the universe, including you and me, radiates a unique energy signature.

If it were theoretically possible to observe the composition of an actual atom with a microscope, what would we see? Imagine a swirling dust devil cutting across the desert's floor. Now remove the sand and dirt from the funnel cloud. What you have left is an invisible, tornado-like vortex. A number of infinitesimally small, dust-devil-like energy vortices called quarks and photons collectively make up the structure of the atom. From far away, the atom would likely appear as a blurry sphere. As its structure came nearer to focus, the atom would become less clear and less distinct. As the surface of the atom drew near, it would disappear. You would see nothing. In fact, as you focused through the entire structure of the atom, all you would observe is a physical void. The atom has no physical structure—the emperor has no clothes!

Remember the atomic models you studied in school, the ones with marbles and ball bearings going around like the solar system? Let's put that picture beside the "physical" structure of the atom discovered by quantum physicists.

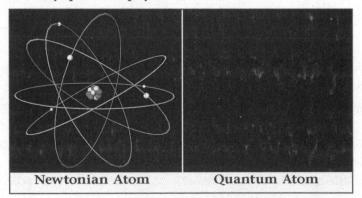

Newtonian Atom   Quantum Atom

No, there has not been a printing mistake; atoms are made out of invisible energy, not tangible matter!

So in our world, material substance (matter) appears out of thin air. Kind of weird, when you think about it. Here you are holding this physical book in your hands. Yet if you were to focus on the book's material substance with an atomic microscope, you would see that you are holding nothing. As it turns out, we undergraduate biology majors were right about one thing—the quantum universe is mind-bending.

Let's look more closely at the "now you see it, now you don't" nature of quantum physics. Matter can simultaneously be defined as a solid (particle) and as an immaterial force field (wave). When scientists study the physical properties of atoms, such as mass and weight, they look and act like physical matter. However, when the same atoms are described in terms of voltage potentials and wavelengths, they exhibit the qualities and properties of energy (waves). (Hackermüller, et al, 2003; Chapman, et al, 1995; Pool 1995) The fact that energy and matter are one and the same is precisely what Einstein recognized when he concluded that $E = mc^2$. Simply stated, this equation reveals that energy (E) = matter (m, mass) multiplied by the speed of light squared (c^2). Einstein revealed that we do not live in a universe with discrete, physical objects separated by dead space. The Universe is *one indivisible, dynamic whole* in which energy and matter are so deeply entangled it is impossible to consider them as independent elements.

They Are Not Side Effects . . . They're Effects!

The awareness that such profoundly different mechanics control the structure and behavior of matter should have offered biomedicine new insights into understanding health and disease. Yet even after the discoveries of quantum physics, biologists and medical students continued to be trained to view the body only as a physical machine that operates in accordance with Newtonian principles. In seeking knowledge of how the body's mechanisms are "controlled," researchers focused their attention on investigating a large variety of physical signals, classified into discrete chemical families, including aforementioned hormones, cytokines, growth factors, tumor suppressors, messengers, and ions. However, because of their Newtonian, materialistic bias, conventional researchers have completely ignored the role that energy vibrations play in health and disease.

In addition, conventional biologists are reductionists who believe that mechanisms of our physical bodies can be understood by taking the cells apart and studying their chemical building blocks. They believe that the biochemical reactions responsible for life are generated through Henry Ford–style assembly lines: one chemical causes a reaction, followed by another reaction with a different chemical, etc. The linear flow of information from A to B to C to D to E is illustrated on the following page.

This reductionist model suggests that if there is a problem in the system, evident as a disease or dysfunction, the source of the problem can be attributed to a malfunction in one of the steps along the chemical assembly line. By providing the cell with a functional replacement part for the faulty element, by prescribing pharmaceutical drugs for example, the defective single point can theoretically be repaired and health can be restored. This assumption spurs the pharmaceutical industry's search for magic-bullet drugs and designer genes.

However, the quantum perspective reveals that the universe is an integration of interdependent energy fields that are entangled in a meshwork of interactions. Biomedical scientists have been particularly confounded because they often do not recognize the massive complexity of the *intercommunication* among the physical parts and the energy fields that make up the whole. The reductionist's perception of a linear flow of information is a characteristic of the Newtonian universe.

In contrast, the flow of information in a quantum universe is *holistic*. Cellular constituents are woven into a complex web of crosstalk, feedback, and feedforward communication loops (see illustration next page). A biological dysfunction may arise from a miscommunication along *any* of the routes of information flow. To adjust the chemistry of this complicated interactive system requires a lot more understanding than just adjusting one of the information pathway's components with a drug. If you change the concentration of C for example, it doesn't just influence the action of D. Via holistic pathways, variations in the concentration of C profoundly influence the behaviors and functions of A, B, and E, as well as D.

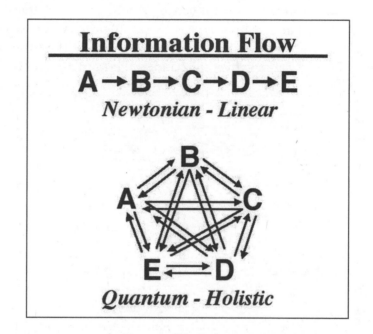

Once I realized the nature of the complex interactions between matter and energy, I knew that a reductionist, linear (A>B>C>D>E) approach could not even come close to giving us an accurate understanding of disease. While quantum physics implied the existence of such interconnected information pathways, recent groundbreaking research in mapping protein-protein interactions in the cell now demonstrates the physical presence of these complex holistic pathways. (Li, et al, 2004; Giot, et al, 2003; Jansen, et al, 2003; Barry 2008) The illustration on page 92 shows the interactions among a few of the proteins in a fruit fly cell. Connecting lines represent protein-protein interactions.

Clearly, biological dysfunctions can result from miscommunication anywhere within these complex pathways. When you change the parameters of a protein at one point in such a complex pathway, you inevitably alter the parameters of other proteins at innumerable points within the entangled networks. In addition, take a look at the seven circles in the illustration below that group proteins according to their physiologic functions. Notice that proteins within one functional group, such as those concerned with sex determination (arrow), also influence proteins with a completely different function, like RNA synthesis (i.e., RNA helicase). Newtonian research scientists have not fully appreciated the extensive interconnectivity among the cell's biological information networks.

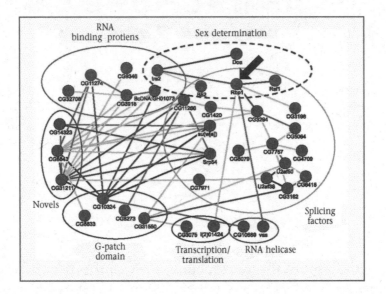

Map of interactions among a very small set of the cellular proteins (shaded and numbered circles) found in a Drosophila (fruit fly) cell. Most of the proteins are associated with the synthesis and metabolism of RNA molecules. Proteins enclosed within ovals are grouped according to specific pathway functions. Connecting lines indicate protein-protein interactions. Protein interconnections among the different pathways reveal how interfering with one protein may produce profound "side effects" upon other related pathways. More widespread "side effects" may be generated when a common protein is utilized in completely different functions. For example, the same Rbp1 protein (arrow) is used in RNA metabolism as well as in pathways associated with sex determination. Reprinted with permission from Science 302:1727-1736. *Copyright 2003 AAAS.*

The mapping of these information network pathways underscores the dangers of prescription drugs. We can now see why pharmaceutical drugs come with information sheets listing voluminous side effects that range from irritating to deadly. When a drug is introduced into the body to treat a malfunction in one protein, that drug inevitably interacts with at least one and possibly many other proteins.

Complicating the drug side-effect issue is also the fact that biological systems are redundant. The same signals or protein molecules may be simultaneously used in different organs and tissues where they provide for completely different behavioral functions. For example, when a drug is prescribed to correct a dysfunction in a signaling pathway of the heart, that drug is delivered by the blood to the entire body. This "cardiac" medicine can unintentionally disturb the function of the nervous system if the brain also uses components of the targeted signaling pathway. While this redundancy complicates the effects of prescription drugs, it is another remarkably efficient result of evolution. Multicellular organisms can survive with far fewer genes than scientists once thought because the same gene products (protein) are used for a variety of functions. This is similar to using only twenty-six letters of the alphabet to construct every word in our language.

In my research on human blood vessel cells, I experienced firsthand the limits imposed by redundant signaling pathways. In the body, histamine is an important chemical signal that initiates the cells' stress response. When histamine is present in the blood that nourishes the arms and legs, the stress signal produces large gaping pores between the cells lining the wall of the blood vessels. The opening of these holes in the blood vessel's wall is the first step in launching a local inflammatory reaction. However, if histamine is added to blood vessels in the brain, the same histamine signal does not cause gaping pores between the lining cells, but instead increases the flow of nutrition to the neurons, enhancing their growth and specialized functions. In times of stress, the increased nutrition signaled by histamine enables the brain to ramp up its activity in order to better deal with the perceived impending emergency. This is an example of how the same histamine signal can create two diametrically opposed effects, depending on the site where the signal is released. (Lipton, et al, 1991)

One of the most ingenious characteristics of the body's sophisticated signaling system is its specificity. If you have a poison ivy rash on your arm, the relentless itchiness results from the release of histamine, the signal molecule that activates an inflammatory response to the ivy's allergen. Since there is no need to start itching all over your body, the histamine is *only* released at the site of the rash. Similarly, when a person is confronted with a stressful life experience, the release of histamine within the brain increases blood flow to the nervous tissues, enhancing the neurological processing required for survival. The release of histamine in the brain to deal with stress behaviors is physically restricted and does not lead to the initiation of inflammation responses in other parts of the body. Like the National Guard, histamine is deployed only where it is needed and for as long as it is needed.

But most of the medical industry's drugs have no such specificity. When you take an antihistamine to deal with the itchiness of an allergic rash, the ingested drug is distributed systemically. It affects histamine receptors wherever they are located throughout the whole body. Yes, the antihistamine will curb the blood vessels' inflammatory response, dramatically reducing allergic symptoms. However, when the antihistamine enters the brain, it inadvertently alters neural circulation that then impacts nerve function. That's why people who take over-the-counter antihistamines may experience allergy relief and also the side effect of feeling drowsy.

A recent example of tragic adverse reactions to drug therapy is the debilitating and life-threatening side effects associated with synthetic hormone replacement therapy (HRT). Estrogen's best-known influence is on the function of the female reproductive system. However, more recent studies on the distribution of estrogen receptors in the body reveal that they, and of course their complementary estrogen signal molecules, play an important role in the normal function of blood vessels, the heart, and the brain. Doctors have routinely prescribed synthetic estrogen to alleviate menopausal symptoms associated with the shutting-down of a woman's reproductive system. However, pharmaceutical estrogen therapy does not focus the drug's effects on the intended target tissues. The drug also impacts and disturbs the estrogen receptors of the heart, the blood vessels, and the nervous system. Synthetic hormone replacement therapy has been shown to have disturbing side effects that result in cardiovascular disease and neural dysfunctions such as strokes. (Shumaker, et al, 2003; Wassertheil-Smoller, et al, 2003; Anderson, et al, 2003; Cauley, et al, 2003; Bath and Gray 2005)

Adverse drug effects, like those contributing to the HRT controversy, are a primary reason why a leading cause of death is iatrogenic illness, i.e., illness resulting from medical treatment. According to conservative estimates published in the *Journal of the American Medical Association,* iatrogenic illness is the third-leading cause of death in this country. More than 120,000 people die from adverse effects of prescribed medications each year. (Starfield 2000) A more recent study, based on the results of a ten-year survey of government statistics, came up with even more dismal figures. (Null, et al, 2003) That study concludes that iatrogenic illness is actually the *leading* cause of death in the United States and that adverse reactions to prescription drugs are responsible for more than 300,000 deaths a year.

These are dismaying statistics, especially for a healing profession that has arrogantly dismissed three thousand years of effective Eastern medicine as unscientific, even though it is based on a deeper understanding of the universe. For thousands of years, long before Western scientists discovered the laws of quantum physics, Asians have honored energy as the principal factor contributing to health and well-being. In Eastern medicine, the body is defined by an elaborate array of energy pathways called meridians. In Chinese physiologic charts of the human body, these energy networks resemble electronic wiring diagrams. Using aids like acupuncture needles, Chinese physicians test their patient's energy circuits in exactly the same manner that electrical engineers "troubleshoot" a printed-circuit board, searching for electrical "pathologies."

Physicians: The Pharmaceutical Patsies

But as admiring as I am of the ancient wisdom of Eastern medicine, I do not want to bash Western doctors who prescribe massive quantities of drugs that contribute to the health profession's lethality.

Medical doctors are caught between an intellectual rock and a corporate hard place; they are pawns in the huge medical industrial complex. Their healing abilities are hobbled by an archaic medical education founded on a Newtonian, matter-only universe. Unfortunately, that philosophy went out of vogue seventy-five years ago, when physicists officially adopted quantum mechanics and recognized that the universe is actually made out of energy.

In their postgraduate years, those same doctors receive their continuing education about pharmaceutical products from drug reps, the errand boys of the corporate healthcare industry. Essentially, these nonprofessionals, whose primary goal is to sell product, provide doctors with "information" about the efficacy of new drugs. Drug companies freely offer this "education" so they can persuade doctors to "push" their products. It is evident that the massive quantities of drugs prescribed in this country violate the Hippocratic Oath taken by all doctors to "First do no harm." We have been programmed by pharmaceutical corporations to become a nation of prescription-drug-popping junkies with tragic results. We need to step back and incorporate the discoveries of quantum physics into biomedicine so that we can create a new, safer system of medicine that is attuned to the laws of nature.

Physics and Medicine: A Day Late and a Dollar Short

The physical sciences have already embraced quantum physics with sensational results. Humanity's wake-up call to the reality of a quantum universe occurred on August 6, 1945. The atomic bomb dropped on Hiroshima that day demonstrated the awesome power of applied quantum theory and dramatically ushered in the Atomic Age. On a more constructive note, quantum physics made possible the electronic miracles that are the foundation of the Information Age. The application of quantum mechanics was directly responsible for the development of TVs, computers, CAT scans, lasers, rocket ships, and cell phones.

But what great and marvelous advances in biomedical sciences can we attribute to the quantum revolution? Let's list them in order of their importance:

It is a very short list—there are hardly any.

Though I stress the need to apply the principles of quantum mechanics in bioscience, I'm not advocating that medicine throw out the valuable lessons it has learned using the principles of Isaac Newton. The newer laws of quantum mechanics do not negate the results of classical physics. The planets are still moving in paths that were predicted by Newton's mathematics.

Quantum physics is a larger realm of awareness that includes and substantially adds to the information provided by Newtonian physics. Consequently, quantum mechanics accounts for what was already known plus a whole new realm of heretofore-unrecognized forces that control the unfolding of our Universe.

A conventional notion regarding the difference between the two physics is that quantum mechanics more specifically applies to molecular and atomic realms while Newtonian laws apply to higher levels of organization, such as organ systems, people, or populations of people. The manifestation of a disease, such as cancer, may show up at a macro level when you can see and feel a tumor. However, the processes that instigated the cancer were initiated at the molecular level within the affected progenitor cells. In fact, most biological dysfunctions (except injuries due to physical trauma) start at the level of a cell's molecules and ions. Hence the need for a biology that integrates both quantum and Newtonian mechanics.

Conventional physics courses suggest that the principles of quantum mechanics that govern wave-particle interactions only apply at the level of atoms. By restricting quantum physics to the subatomic world, it has become a general assumption that quantum mechanisms do not apply to our personal lives and world affairs. Therefore, today's physicists have completely failed to inform the public of the purely mental nature of the Universe.

Fortunately, leaders in the field, such as Johns Hopkins University physicist Richard Conn Henry, are addressing the misperceptions about the perceived primacy of the material world. Henry offered an elegantly simple definition on the true nature of the Universe: "The Universe is immaterial—mental and spiritual. Live, and enjoy." (Henry 2005) Simply, the mechanics of quantum physics applies at every level of the Universe from the Big Bang down to the quarks in atoms.

There have, thankfully, been some visionary biologists who have advocated for the integration of Newtonian and quantum physics. More than forty years ago the renowned Nobel Prize–winning physiologist Albert Szent-Györgyi published a book called *Introduction to a Submolecular Biology.* (Szent-Györgyi 1960) His text was a noble effort to educate the community of life scientists about the importance of quantum physics in biological systems. Unfortunately, his traditional peers, who considered the book to be the ravings of a once brilliant but now senile old man, merely lamented the "loss" of their former colleague. Biologists in the main have still not recognized the importance of Szent-Györgyi's book, but research suggests that sooner or later they will have to because the weight of scientific evidence is toppling the old materialist paradigm. You recall the movements of protein molecules that are the stuff of life? Scientists have tried to predict those movements using the principles of Newtonian physics, to no avail. By now, I bet you can guess why: in 2000, an article by V. Pophristic and L. Goodman in the journal *Nature* revealed that the laws of quantum physics, not Newtonian laws, control a molecule's life-generating movements. (Pophristic and Goodman 2001)

Reviewing this groundbreaking study for *Nature*, biophysicist F. Weinhold concluded: "When will chemistry textbooks begin to serve as aids, rather than barriers, to this enriched quantum-mechanic perspective on how molecular turnstiles work?" He further emphasized: "What are the forces that control the twisting and folding of molecules into complex shapes? Don't look for the answers in your organic chemistry textbook." (Weinhold 2001) Yet organic chemistry provides the mechanistic foundation for biomedicine; and as Weinhold notes, that branch of science is so far out of date that its textbooks have yet to recognize quantum mechanics. Conventional medical researchers have no understanding of the molecular mechanisms that truly provide for life.

Frontier research on the mechanisms that cause proteins to change shape reveals the primacy of quantum properties in producing the movements that result in life. These studies demonstrate that the manipulation of the quantum properties of matter can influence the course of biochemical reactions. (Schulten 2000; Chergui 2006; Gaidos 2009)

Hundreds upon hundreds of other scientific studies over the last fifty years have consistently revealed that "invisible forces" of the electromagnetic spectrum profoundly impact every facet of biological regulation. These energies include microwaves, radio frequencies, the visible light spectrum, extremely low frequencies, acoustic frequencies, and even a newly recognized form of force known as scalar energy. Specific frequencies and patterns of electromagnetic radiation regulate DNA, RNA, and protein syntheses; alter protein shape and function; and control gene regulation, cell division, cell differentiation, morphogenesis (the process by which cells assemble into organs and tissues), hormone secretion, and nerve growth and function. Each one of these cellular activities is a fundamental behavior that contributes to the unfolding of life. Though these research studies have been published in some of the most respected mainstream biomedical journals, as of 2010 their revolutionary findings have not been incorporated into the medical school curriculum. (Liboff 2004; Goodman and Blank 2002; Sivitz 2000; Jin, et al, 2000; Blackman, et al, 1993; Rosen 1992, Blank 1992; Tsong 1989; Yen-Patton, et al, 1988)

An important study forty years ago by Oxford University biophysicist C. W. F. McClare calculated and compared the efficiency of information transfer between energy signals and chemical signals in biological systems. His research, "Resonance in Bioenergetics," published in the *Annals of the New York Academy of Science,* revealed that energetic signaling mechanisms such as electromagnetic frequencies are a hundred times more efficient in relaying environmental information than physical signals such as hormones, neurotransmitters, growth factors, etc. (McClare 1974)

It is not surprising that energetic signals are so much more efficient. In physical molecules, the information that can be carried is directly linked to a molecule's available energy. However, the chemical coupling employed to transfer their information is accompanied by a massive loss of energy due to the heat generated in making and breaking chemical bonds. Because thermo-chemical coupling wastes most of the molecule's energy, the small amount of energy that remains limits the amount of information that can be carried as the signal.

We know that living organisms must receive and interpret environmental signals in order to stay alive. In fact, survival is directly related to the speed and efficiency of signal transfer. The speed of electromagnetic energy signals is 186,000 miles per second, while the speed of a diffusible chemical is considerably less than one centimeter per second. Energy signals are a hundred times more efficient and infinitely faster than physical chemical signaling. What kind of signaling would your trillion-celled community prefer? Do the math!

Buying the Pharm

I believe the major reason why energy research has been all but ignored comes down to dollars and cents. The trillion-dollar pharmaceutical industry puts its research money into the search for magic bullets in the form of chemicals because pills mean money. If energy healing could be made into tablet form, drug manufacturers would get interested quickly.

Instead, they identify deviations in physiology and behavior that vary from some hypothetical norm as unique disorders or dysfunctions, and then they educate the public about the dangers of these menacing disorders. Of course, the oversimplified symptomology used in defining the dysfunctions prevalent in drug company advertisements has viewers convinced they are afflicted by that particular malady. "Do you worry? Worry is a primary symptom of a 'medical condition' called anxiety disorder. Stop your worry. Tell your doctor you want Addictazac, the new passion-pink drug."

Meanwhile, the media essentially avoids the issue of deaths by medicine by directing our attention to the dangers of illicit drugs. They admonish us that using drugs to escape life's problems is not the way to resolve one's issue. Funny . . . I was just going to use that exact sentence to describe my concerns about the overuse of legal drugs. Are they dangerous? Ask the people who died last year. Using prescription drugs to silence a body's symptoms enables us to ignore personal involvement we may have with the onset of those symptoms. The overuse of prescription drugs provides a vacation from personal responsibility.

Our drug mania reminds me of a job at an auto dealership I held while in graduate school. At 4:30 on a Friday afternoon, an irate woman came into the shop. Her car's "service engine" light was flashing, even though her car had already been repaired for that same problem several times. At 4:30 on a Friday afternoon, who wants to work on a balky problem and deal with a furious customer? Everyone was quiet, except for one mechanic who said, "I'll take care of it." He drove the car back into the bay, got in behind the dashboard, removed the bulb from the signal light and threw it away. Then he opened a can of soda and lit a cigarette. After a suitable time, during which the customer thought he was actually fixing the car, the mechanic returned and told the woman her car was ready. Thrilled to see that the warning light had stopped flashing, she happily drove off into the sunset. Though the cause of the problem was still present, the symptom was gone. Similarly, pharmaceutical drugs suppress the body's symptoms but most never address the cause of the problem.

"Wait," you say. "Times have changed." We are now more educated to the dangers of drugs and more open to alternative therapies. It is true that because half of Americans visit complementary health practitioners, traditional doctors can no longer put their heads in the sand and hope other approaches go away. A few insurance companies have even started to pay for services they once deemed quackery, and major teaching hospitals allow a limited number of such practitioners inside.

But even today very little scientific rigor has been marshaled to assess the effectiveness of complementary medicine. The National Institutes of Health did create an "alternative medicine" branch, thanks to pressure from the public. But, in my opinion, that is only a token gesture to quell activists and consumers who spend lots of money on alternative health care. There are no serious research funds available for studying energy medicine. The rub is that without supportive research, energy-based healing modalities are officially labeled "unscientific."

Good Vibes, Bad Vibes, and the Language of Energy

Though conventional medicine still has not focused on the role energy plays as "information" in biological systems, ironically, it has embraced noninvasive scanning technologies, which read such energy fields. Quantum physicists have created energy-scanning devices that can analyze the frequencies emitted by specific chemicals. These scanning systems enable scientists to identify the molecular composition of materials and objects. Physicists have adapted these devices to read the energy spectra emitted by our body's tissues and organs. Because energy fields travel easily through the physical body, these modern devices, such as CAT scans, MRIs, and positron emission tomography (PET) scans, can detect disease noninvasively. Physicians are able to diagnose internal problems by differentiating the spectral energy character of healthy and diseased tissue in the scanned images.

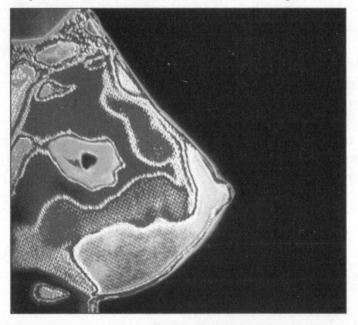

Mammogram. Note the above illustration is not a photograph of a breast, it is an electronic image created from scanning the radiant energy characteristics of the organ's cells and tissues. Differentials in the energy spectra enable radiologists to distinguish between healthy and diseased tissues (the black spot in the center).

The energy scan illustrated on page 102 reveals the presence of breast cancer. The diseased tissue emits its own unique energy signature, which differs from the energy emitted by surrounding healthy cells. The energy signatures that pass through our bodies travel through space as invisible waves that resemble ripples on a pond. If you drop a pebble into a pond, the "energy" carried in the falling pebble (due to the force of gravity pulling on its mass) is transmitted to the water. The ripples generated by the pebble are actually energy waves passing through the water.

If more than one pebble is thrown into the water at the same time, the spreading ripples (energy waves) from each source can interfere with each other, forming composite waves where two or more ripples converge. That interference can be either constructive (energy-amplifying) or destructive (energy-deflating).

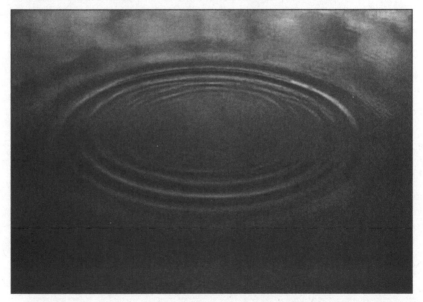

Dropping two pebbles of the same size, from the same height, and at exactly the same time, coordinates the wave action of their ripples. The ripples from each pebble converge on each other. Where the ripples overlap, the combined power of the interacting waves is doubled, a phenomenon referred to as constructive inter- ference, or *harmonic resonance*. When the dropping of the pebbles is not coordinated, their energy waves are out of sync. As one wave is going up, the other is going down. At the point of convergence these out of sync energy waves cancel each other. Instead of a doubling of the energy where the ripples interfere with each other, the water is calm . . . there is no energy wave. This phenomenon of canceling energy waves is called destructive interference.

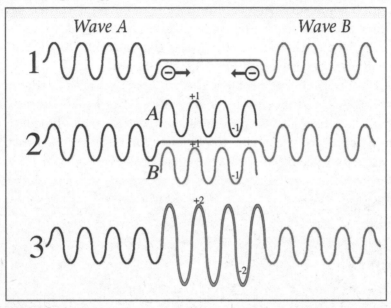

Constructive Interference. In 1 above, two sets of ripples are moving across the surface of water toward each other. As illustrated, both wave A and B are moving toward each other with their ripples in phase, in this case both waves are going up and down at the same time. Their cycle patterns are aligned. The waves merge together at the interface where two ripples meet. To illustrate the consequence of this merger, the waves are drawn with one above the other in figure 2. Where the amplitude of A is +1, the amplitude of B is also +1. Add the two together, and the resulting amplitude of the composite wave at that point is +2. Likewise, where A is –1 so is B, together the total amplitude will be –2. The resulting higher amplitude composite wave is illustrated in 3.

The behavior of energy waves is important for biomedicine because vibrational frequencies can alter the physical and chemical properties of an atom as surely as physical signals like histamine and estrogen. Because atoms are in constant motion, which you can measure by their vibration, they create wave patterns similar to the expanding ripples from the thrown pebbles we talked about above. Each atom is unique because the distribution of its negative and positive charges, coupled with its spin rate, generates a specific vibration or frequency pattern. (Oschman 2000)

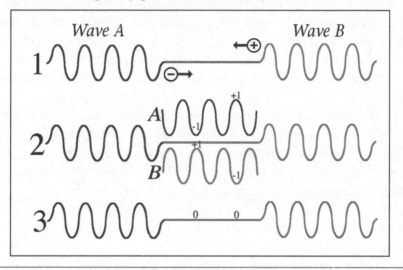

Destructive Interference. In figure 1, the ripples derived from first pebble, labeled as Wave A, are moving from left to right. Wave B, moving right to left, represents the ripples from a second pebble dropped shortly after the first. Since the pebbles did not hit the water at the same time, the waves will not be aligned when they merge at the interface, they will be "out of phase." In the illustration, Wave A is leading with a negative amplitude, and Wave B is leading with a positive amplitude. Where they meet in figure 2, the waves are mirror-images of each other, the high amplitude (+1) of one wave is aligned with the low amplitude (-1) of the other, and vice versa. As shown in 3, the amplitude values of each wave cancel each other out, so that the composite wave having 0 amplitude is no wave at all . . . it's flat!

Scientists have devised a way to stop an atom dead in its tracks by exploiting its energy waves. They first identify the frequency of a specific atom and then tune a laser to emit the same frequency. Though the atom and the photoelectric frequency emit the same wave pattern, the laser's waves are designed to be out of sync with those of the atom. When the light wave interacts with the atom's wave, the resulting destructive interference cancels the atom's vibrations and it stops spinning. (Chu 2002; Rumbles 2001)

To increase an atom's vibration rather than stop its movement, researchers select light waves with vibrations that are harmonically resonant and in phase with those of the atom. The vibrations can be of electromagnetic or acoustic origin. When, for example, a skilled vocalist like Ella Fitzgerald maintains a note that is harmonically resonant with the atoms of a crystal goblet, the goblet's atoms absorb her sound waves. Through the mechanics of constructive interference, the added energy of resonant sound waves causes the goblet's atoms to vibrate faster. Eventually the atoms absorb so much energy that they vibrate fast enough to break free from the bonds that hold them together. When that happens, the goblet actually explodes.

Doctors use constructive interference mechanics to treat kidney stones, a rare case where the laws of quantum physics have been harnessed as a therapeutic tool in modern medicine. Kidney stones are crystals whose atoms vibrate at a specific frequency. Doctors noninvasively focus a harmonic frequency on the kidney stone. Constructive interference results when the focused energy waves interact with the atoms in the kidney stones. Like the atoms in the crystal goblet example above, the atoms of the kidney stones vibrate so quickly that the stones explode and dissolve. The small, remaining fragments can then be easily passed from the system without the excruciating pain that accompanies large, unexploded stones.

The science of physics implies that the same harmonic resonance mechanism, by which sound waves destroy a goblet or a kidney stone, can enable similar energy harmonics to influence the functions of our body's chemistry. But biologists have not explored these mechanisms with the passion with which they pursue new drugs. That is unfortunate because there is enough scientific evidence to suspect that we can tailor a waveform as a therapeutic agent in much the same way we now modulate chemical structures with drugs.

There was a time in medicine when electrotherapy was used extensively. At the end of the nineteenth century, the development of batteries and other devices that produce electromagnetic fields led to hastily constructed machines that were supposed to cure disease. The public sought out practitioners of this new-fangled healing art called radioesthesia. Word spread that these devices were very effective. In fact, they became so popular that magazines were likely to tout ads that read something like: "Be a Radioesthesiast! Only $9.99—includes instructions!" By 1894, over 10,000

U.S. physicians, as well as an untold number of self-trained home consumers, were regularly using electrotherapy.

In 1895, D. D. Palmer created the science of chiropractic. Palmer recognized that the flow of energy through the nervous system is critical to health. He focused on the mechanics of the vertebral column, the conduit through which spinal nerves provide information to the body. He developed skills to assess and tune the flow of information by adjusting the backbone's tensions and pressures.

The medical profession became threatened by Palmer's chiropractors as well as homeopathic healers, radioesthesiasts, and other drugless practitioners who were taking away much of their business. The Carnegie Foundation published the Flexner Report in 1910 that called for all medical practices to be based on proven science. Because physicists had not yet discovered the quantum universe, energy medicine was incomprehensible to science. Denounced by the American Medical Association, chiropractic and other energy-based modalities fell into disrepute. Radioesthesiasts disappeared completely.

In the last forty years, chiropractic has made great inroads in the healing arts. In 1990, chiropractors won a lengthy court battle against the medical monopoly when the American Medical Association was found guilty of illegal attempts to destroy the profession. Since then, chiropractic has spread its sphere of influence—it is even accepted in some hospitals. And despite electrotherapy's checkered past, neuroscientists are conducting exciting new research in the area of vibrational energy therapies.

The brain has long been recognized to be an electrical organ, which is why electroshock therapy has historically been used to treat depression. But scientists are now working on less invasive tools to treat the electric brain. A recent article in *Science* touted the beneficial effects of transcranial magnetic stimulation (TMS), which stimulates the brain with magnetic fields. (Helmuth 2001; Hallet 2000) TMS is an updated version of the same nineteenth century radioesthesia healing techniques that were once denounced by conventional medicine. New studies suggest that TMS can be a powerful therapeutic tool. If used properly, it can ease depression and alter cognition.

It is clear that we need interdisciplinary research in this promising and understudied area, research that encompasses quantum physics, electrical engineering, and chemistry, as well as biology. Such research will be particularly welcomed because it is likely to result in therapies with far fewer side effects than drugs. But the research will only confirm what scientists and nonscientists already "know" but may not realize they know: all organisms, including humans, communicate and read their environment by evaluating energy fields. Because humans are so dependent on spoken and written language, we have neglected our energy-sensing communication system. As with any biological function, a lack of use leads to atrophy. Interestingly, aborigines still utilize this hyper-sensory capacity in their daily lives. For them there has been no "sensory" atrophy. For example, Australian aborigines can sense water buried deep beneath the sand, and Amazonian shamans communicate with the energies of their medicinal plants.

You no doubt on occasion get a glimmer of your ancient sensing mechanism. Have you ever walked down a dark street at night and instantly felt drained of energy? What were you experiencing? Destructive interference, just like out-of-sync pebbles thrown into a pond, or, in popular jargon, bad vibes! Remember unexpectedly meeting that special someone in your life and becoming so energized you felt "high"? You were experiencing constructive interference, or good vibes.

When I gave up my view that we are inert matter, I realized not only that the science of my chosen career was out of date but also that I needed to promote more constructive interference in my own life. I needed a personal quantum-physics-inspired tune-up! Rather than focusing on creating harmonic energies in my life, I was going through life willy-nilly, mindlessly expending energy. That is the equivalent of heating a house in the dead of winter while leaving the doors and windows open. I started closing those doors and windows by carefully examining where I was wasting my energy. It was easy for me to close some of them. For example, it was easy to get rid of energy-draining activities like those deadly faculty parties. It was harder to get rid of the energy-draining defeatist thinking in which I habitually engaged. Thoughts consume energy as surely as does marathon running, as we'll see in the next chapter.

I needed a quantum tune-up. And so, I've made clear, does bio-medicine. But as I said earlier, we are already in the midst of a very slow shift in medicine, propelled by consumers who are seeking out complementary medicine practitioners in record numbers. It's been a long time coming, but the quantum biological revolution is nigh. The medical establishment will eventually be dragged, half kicking and screaming, full force into the quantum revolution.

* * *

The quantum biological revolution that I said was "nigh" a decade ago is in full swing (and, as I explained in the Prologue, I've made lots of progress on my personal quantum tune-up as well!). New research on various biophysical fronts has convinced an increasing number of biologists that there is quantum magic behind cell signaling, protein behavior, and even the origins of life.

Specifically, recent studies of protein behavior point to the influence of a variety of quantum mechanisms that shape biological behavior, including energy entanglement (wherein one energy source couples and influences another energy source), tunneling (wherein particles pass through physical barriers), and superposition (wherein particles *simultaneously* experience all possible pathways and then choose the most effective one on which to travel—these particles are effectively in many places at the same time!). (Sarovar, et al, 2010)

The scientists studying the intersection of these counterintuitive, weird phenomena and biology have created a quantum beachhead on territory once solely owned by classical Newtonian biologists. That beachhead includes the European Science Foundation, which in 2011 established Farquest, an initiative exclusively devoted to creating collaborative efforts for assessing the role of quantum information, especially in biological systems. And the inherent significance of new quantum biophysics led DARPA, the U.S. defense research agency, to create a nationwide quantum biology network in 2010 to explore this burgeoning area of research.

This expanding network is producing compelling research that challenges "facts" that biologists like me memorized in school—for example, the "fact" that signals controlling cell behavior and genetics are carried only in the substance of chemistry, such as hormones, drugs, atoms, and ions (e.g., Ca^+, Na^+, and K^+). That notion was upended by experiments reported by Chaban in 2013 that revealed that nerve cells *outside* physical barriers influence the activity of nerve cells within sealed chambers. When healthy nerve cells surround the barrier, the encapsulated nerve cells express a normal calcium signal process, but when cancerous or dying cells surround the physical barrier, enclosed nerve cells process calcium signals in a completely different manner. Because the barrier prevents *physical* signals from influencing cell behavior, the nerve cells must be communicating with one another across the barrier using a *nonphysical, energetic* signaling mechanism. (Chaban, et al, 2013)

Researchers are also discovering nonorthodox communication channels in the plant world. Plant ecologists have long known that neighboring species, when planted in close proximity, have a positive or negative impact on one another through competitive or cooperative interactions. Neighbor plants can influence a seed's germination timing and its eventual success. A seedling's ability to engage with its neighbors is advantageous because it enables the plant to regulate its genetics and behaviors to best adapt to the surrounding environment. A number of studies have established that this communication among plants exists through three channels: light, physical touch, and chemicals. But the newest research has concluded that plant communication is mediated by some nonconventional and nonphysical (energetic) method as well.

For example, in a recent study, Australian researchers, taking their cue from gardeners who believe that planting basil near chili seeds ensures that they'll have lots of spicy food, divided 3,600 chili seeds into three groups, studied each group's germination rates, and then repeated the experiment with 3,600 chili plants they used as a control group. The results of the experiments corroborated gardening folk wisdom: the presence of basil "enhanced germination rates" of the seeds. But the most interesting result was that this enhancement occurred even when the three signals long studied by scientists (light, chemicals, and physical touch) were blocked. The researchers concluded that those three signaling mechanisms are "clearly not necessary for chili seeds and basil plants to sense each other's presence." The authors of the study state that "no mechanistic explanation of how plants may perform the observed feat is yet available." But they go on to explain that such a communications channel requires "the emission of a signal that not only propagates rapidly to convey real-time information about neighbouring plants but also can be analyzed quickly." (Gagliano and Renton 2013) Sounds like quantum, energetic communication to me!

Quantum magic is also being uncovered in studies of one of life's most important biological activities: photosynthesis. Photosynthesis involves the capture of a photon of light by chlorophyll, a multiprotein complex, and how the energy of that light is used to mobilize electrons that empower the creation of organic chemistry from inorganic CO_2 and water. The electrons transferred in the photosynthesis process can choose among many different pathways as they travel through the protein complex. However, because photosynthesis is a highly efficient process, electrons ultimately choose to utilize a single pathway. But how? Quantum mechanics, in the form of superposition, enables a quantum particle, an electron in this case, to *simultaneously* experience all possible pathways and then choose the most effective one on which to travel. (Richards, et al, 2012)

The photosynthesis process requires a specific and precise alignment of the molecular components in the complex in order to effectively steer this reaction to its successful conclusion. In 2006, Canadian physicists and chemists demonstrated that manipulating vibrational frequencies and exploiting the wave-like nature of matter can selectively direct the behavior of atomic and molecular systems. (Prokhorenko, et al, 2006) But chlorophyll proteins work at relatively high ambient temperatures that cause these molecules to experience random thermal vibrations. How are electrons efficiently shuttled through the chlorophyll complex in the presence of these random vibrations? Combining molecular dynamics and quantum chemistry to study electron transfer processes, physicists at the University of California, San Diego offered a profound solution to this problem. Their study revealed that electron transfer occurs through a web of quantum tunneling pathways created through constructive or destructive interference patterns typical of the wave-like processes of quantum mechanics I described in this chapter. (Balabin and Onuchic 2000)

The challenge with all this new research is that while it is an indisputable fact that immaterial, energetic signals can control biology, there is no conventional known mechanism to explain such phenomena. Well, at least there is no known classical (Newtonian) mechanism to account for this behavior. In our current world, information must be measurable before it becomes real or accepted by the scientific community, and a major problem in studying the biological influence of quantum signaling is that it is difficult to measure this energy.

Frequently, this form of communication is so subtle that the amplitude of life-influencing signals is way below the resolution of scientific instrumentation. In 2014, an international research team illuminated the fact that cellular signaling by the Ras family of membrane proteins, one of the most important components of signaling networks in biology, is so subtle that it cannot be distinguished from background "noise" that experimenters conventionally disregard as a technical artifact. (Iversen, et al, 2014)

Energy fields influencing cell membrane receptors, such as those controlling Ras proteins, represent the interface wherein environmental signals control cell functions. Experimental manipulation of environmental EMF (electromagnetic field) frequencies has been shown to profoundly influence the activity of the cell's sodium ($Na+$), potassium ($K+ATPase$), and calcium ($Ca2+ATPase$) ion protein channels. (Guan and Reed 2012) Since these membrane proteins control the cell's electrical activities, including the development and maintenance of the cell's membrane potential, environmental electromagnetic fields can shape the health and fate of biological systems. For example, it is now well established that microwave radiation, associated with cell phones and other electronic devices, interferes with and disrupts normal cell behaviors and can possibly lead to dysfunction and disease. (Kesari, et al, 2013)

Of course, as always in science, there were a number of quantum biology pioneers whose prescient work was dismissed. In the late 1930s, Harold Saxton Burr, a Professor of Anatomy at the Yale

University School of Medicine, sought to measure and characterize the immaterial "bio-magnetic field" associated with living organisms. Burr steadfastly believed that life not only exhibited electromagnetic properties but that these same properties represented an organizing principle that shaped the growth and development of cells, tissues, and organs. His research in 1938, employing the then state-of-the-art electrical measurements, provided indisputable evidence for his belief. (Burr and Northrop 1939)

At the same time, American inventor Royal Rife independently verified and advanced Burr's hypothesis. Rife created a "beam" machine that would weaken or destroy pathogens and cancer cells by focusing specific constructive or destructive interference energy fields on their cytoplasmic chemistry. (Valone 2000) The experiments of both Burr and Rife revealed that the appearance of certain diseases, such as cancer, is preceded by a measurable change in an organism's energy field. Most importantly, by simply altering the frequencies of the cell's energy field, they could ameliorate these pathologic disturbances. The scientific focus on describing the "chemical" nature of living organisms led conventional materialist scientists to bury the thought-provoking findings of Burr, Rife, and many others that show that life is intimately connected to invisible energy fields.

Recent studies are harder to ignore because, though they demolish long-held assumptions, they also use convincing state-of-the-art technology, including advanced genetic engineering techniques, new microscopes with super-high resolution and sensitivity that provide images of living cells, and fluorescent markers that scientists can attach to proteins to clearly delineate their movements. These instruments are not only providing insight into how quantum mechanics affects everyday biology, but also are motivating researchers to better understand how biological systems use quantum mechanisms to develop new, efficient technologies for organic syntheses and energy capture (e.g., solar cells). (Arndt, et al, 2009)

One bioengineering innovation known as optogenetics, a technology by which cells can be programmed to respond to specific environmental frequencies, makes it clear that cellular molecules, especially proteins, represent physical nano-devices whose behavior can be controlled by impacting their inherent electric and magnetic properties using applied environmental energy fields. In optogenetics, engineered DNA blueprints that combine a light-sensitive pigment molecule with a specific function-controlling cell membrane receptor are introduced into nerve cells via a virus. The infected DNA then encodes the engineered receptor protein that is naturally incorporated into the cell's membrane. When exposed to a specific frequency of light, the new receptors on these infected cells are activated, which in turn engages the cell's function. (Fenno, et al, 2011)

Despite this groundbreaking work in quantum biophysics and thanks in part to the vast resources of the pharmaceutical industry, an entity that thrives on sales of chemical signals and abhors drug-free energy medicine, the conventional world of medicine is still essentially ignoring the role of energy in shaping biological expression. When I first wrote this chapter ten years ago I said that though the medical establishment would eventually join the quantum revolution, it would have to be "dragged, half kicking and screaming." Sadly, they're still kicking and screaming. In his lecture at the 2004 Lindau Nobel Laureate Meeting, Nobel Prize–winning University of Cambridge physicist Brian Josephson said that the scientific establishment suffers from "Pathological Disbelief," a condition he described as "I wouldn't believe it even if it were true." At the beginning of the lecture, he included the following admonition: "WARNING: Readers may find some of the ideas in this lecture disturbing; they may conflict with various deeply held beliefs." Josephson recalled how the topics of meteorites and continental drift were originally and vehemently dismissed by scientists as impossible phenomena. After a long period of time and the acquisition of abundant evidence, science finally caved and accepted the impossible as real. (Josephson 2004)

Josephson has also questioned today's negative scientific attitude about the validity of homeopathy, a more than 200-year-old alternative medicine system that is routinely dismissed by the medical establishment as quackery. The fact that homeopathy is favored by the British royal family hasn't stopped the British Medical Association from describing it as "witchcraft." When asked by *New Scientist* editors how he became an advocate of unconventional medicine, Josephson said it was when he saw French immunologist Jacques Benveniste present his research at a conference. (Benveniste published the first research article to validate homeopathy in the prestigious journal *Nature*.) Josephson noted that Benveniste's presentation "provoked irrationally strong reactions from scientists." He adds: "I was struck by how badly he was treated." (George 2006) Another Nobel Laureate, Luc Montagnier, the French virologist who won the Prize in 2008 for discovering the AIDS virus, has studied homeopathy and lauded Benveniste as a "modern Galileo."

Benveniste was attacked and demonized because he investigated a subject that the orthodox medical and scientific community had dismissed as a metaphysical oddity. Because homeopathy relies on highly diluted remedies, conventional scientists have disregarded those remedies out of hand, arguing that the homeopathic solutions are so dilute that they no longer contain any of the original signal molecules. In response to this critique on dilution, Montagnier told *Science*: "High dilutions of something are not nothing. They are water structures which mimic the original molecules." (Enserink 2010)

Additionally, what Josephson calls the "simple-minded assessment" that fluid water molecules cannot have a structure to maintain information is now invalid. This limited thinking does not include new insights offered by the science of liquid crystals that reveal how a flowing fluid, such as water, can maintain an ordered structure over macroscopic distances. These new insights nullify the standard refutations of homeopathy put forward by naysayers.

While Big Pharma and closed minds prevent some scientists from jumping on the quantum bandwagon, the public leads the way into the frontiers of energy medicine. A National Health Interview Survey found that 83 million American adults spent $38 billion on complementary and alternative healthcare in the twelve months before the survey was published in 2007. While there are no comparable more recent statistics, anecdotal evidence points to the public's continuing enthusiasm for remedies outside the traditional biomedical model. (Nahin, et al, 2009) For example, the science of how transcranial magnetic stimulation (TMS) alters cognition, reported in the first edition of this book, is now being employed by lay audience enthusiasts who are building or buying electromagnetic field stimulators to enhance brain function. The new version of this technology, referred to as *transcranial direct current stimulation* (tDCS), sends small amounts of stable electric current into the scalp. Depending on the region of the brain being stimulated, tDCS can influence neural activity that results in increasing attention, expanding memory, improving visual abilities and mathematical skills, and alleviating symptoms of depression. (Sanders 2014)

The public's use of electrical stimuli to influence brain functions, a modern version of radioesthesia, is still relatively fringe, yet it is a growing field of interest as demonstrated by increasing participation in an online "do-it-yourself zapper" brain-hacking message board moderated by twenty-two-year-old Nathan Whitmore, a neuroscience researcher at the National Institute on Aging and advocate of DIY brain zapping.

The quantum biophysics revolution is well underway. And yes . . . it may be televised!

CHAPTER 5

BIOLOGY and BELIEF

In 1952 a young British physician made a mistake. It was a mistake that was to bring short-lived scientific glory to Dr. Albert Mason. Mason tried to treat a fifteen-year-old boy's warts using hypnosis. Mason and other doctors had successfully used hypnosis to get rid of warts, but this was an especially tough case. The boy's leathery skin looked more like an elephant's hide than a human's, except for his chest, which had normal skin.

Mason's first hypnosis session focused on one arm. When the boy was in a hypnotic trance, Mason told him that the skin on that arm would heal and turn into healthy, pink skin. When the boy came back a week later, Mason was gratified to see that the arm looked healthy. But when Mason brought the boy to the referring surgeon, who had unsuccessfully tried to help the boy with skin grafts, he learned that he had made a medical error. The surgeon's eyes were wide with astonishment when he saw the boy's arm. It was then that he told Mason that the boy was suffering, not from warts, but from a lethal genetic disease called congenital ichthyosis. By reversing the symptoms using "only" the power of the mind, Mason and the boy had accomplished what had until that time been considered impossible. Mason continued the hypnosis sessions, with the stunning result that most of the boy's skin came to look like the healthy, pink arm after the first hypnosis session. The boy, who had been mercilessly teased in school because of his grotesque-looking skin, went on to lead a normal life.

When Mason wrote about his startling treatment for ichthyosis in the British Medical Journal in 1952, his article created a sensation. (Mason 1952) Mason was touted in the media and became a magnet for patients suffering from the rare, lethal disease that no one before had ever cured. But hypnosis was in the end not a cure-all. Mason tried it on a number of other ichthyosis patients, but he was never able to replicate the results he had had with the young boy. Mason attributes his failure to his own belief about the treatment. When Mason treated the new patients he couldn't replicate his cocky attitude as a young physician thinking he was treating a bad case of warts. After that first patient, Mason was fully aware that he was treating what everyone in the medical establishment knew to be a congenital, "incurable" disease. Mason tried to pretend that he was upbeat about the prognosis, but he told the Discovery Health Channel, "I was acting." (Discovery Health Channel 2003)

How is it possible that the mind can override genetic programming, as it did in the case above? And how could Mason's *belief* about that treatment affect its outcome? The New Biology suggests some answers to those questions. We saw in the last chapter that matter and energy are entangled. The logical corollary is that the mind (energy) and body (matter) are similarly bound, though Western medicine has tried valiantly to separate them for hundreds of years.

In the seventeenth century, René Descartes dismissed the idea that the mind influences the physical character of the body. Descartes' notion was that the physical body was made out of matter and the mind was made out of an unidentified but clearly immaterial substance. Because he couldn't identify the nature of the mind, Descartes left behind an irresolvable philosophical conundrum: since only matter can affect matter, how can an immaterial mind be "connected" to a material body? The nonphysical mind envisioned by Descartes was popularly defined as the "Ghost in the Machine" by Gilbert Ryle sixty years ago in his book *The Concept of Mind*. (Ryle 1949) Traditional biomedicine, whose science is based on a Newtonian matter-only universe, embraced Descartes' separation of mind and body. Medically speaking, it would be far easier to fix a mechanical body without having to deal with its meddling "ghost."

The reality of a quantum universe reconnects what Descartes took apart. Yes, the mind (energy) arises from the physical body, just as Descartes thought. However, our new understanding of the universe's mechanics shows us how the physical body can be affected by the immaterial mind. Thoughts, the mind's energy, directly influence how the physical brain controls the body's physiology. Thought "energy" can activate or inhibit the cell's function-producing proteins via the mechanics of constructive and destructive interference, described in the previous chapter. That is why, when I took the first step toward changing my life, I actively monitored where I was expending my brain's energy. I had to examine the consequences of energy I invested in my thoughts as closely as I examined the expenditures of energy I used to power my physical body.

Despite the discoveries of quantum physics, the mind-body split in Western medicine still prevails. Scientists have been trained to dismiss cases like the boy above, who used his mind to heal a genetically "mandated" disease, as quirky anomalies. I believe, on the contrary, that scientists should embrace the study of these anomalies. Buried in exceptional cases are the roots of a more powerful understanding of the nature of life—"more powerful" because the principles behind these exceptions trump established "truths." The fact is that harnessing the power of your mind can be *more* effective than the drugs you have been programmed to believe you need. The research I discussed in the last chapter found that energy is a *more* efficient means of affecting matter than chemicals.

Unfortunately, scientists most often deny rather than embrace exceptions. My favorite example of scientific denial of the reality of mind-body interactions relates to an article that appeared in *Science* about nineteenth-century German physician Robert Koch, who along with Pasteur founded the Germ Theory. The Germ Theory holds that bacteria and viruses are the primary cause of disease. A modified version of that theory is widely accepted now, but in Koch's day it was more controversial. One of Koch's critics was so convinced that the Germ Theory was wrong that he brazenly wolfed down a glass of water laced with *vibrio cholerae*, the bacterium Koch believed caused cholera. To everyone's astonishment, the man was completely unaffected by the virulent pathogen. The *Science* article published in 2000 describing the incident stated: "For unexplained reasons he remained symptom free, but nevertheless incorrect." (DiRita 2000)

The man survived and *Science,* reflecting the unanimity of opinion on the Germ Theory, had the audacity to say his criticism was *incorrect?* If it is claimed that this bacterium is the cause of cholera and the man demonstrates that he is unaffected by the germs . . . how can *he* be "incorrect"? Instead of trying to figure out how the man avoided the dreaded disease, scientists blithely dismiss this and other embarrassing "messy" exceptions that spoil their theories. Remember the "dogma" that genes control biology? Here is another example in which scientists, bent on establishing the validity of *their* truth, ignore pesky exceptions. The problem is that there *cannot* be exceptions to a theory; exceptions simply mean that a theory is not fully correct.

A current example of a reality that challenges the established beliefs of science concerns the ancient religious practice of fire-walking. Seekers gather together daily to stretch the realms of conventional awareness by walking across beds of hot coals. Measurement of the stone's temperature and duration of exposure are enough to cause medically relevant burns on the feet, yet thousands of participants emerge from the process totally unscathed. Before you jump to the conclusion that the coals were not really that hot, consider the numbers of participants who get seriously scalded walking across the same bed of coals.

Similarly, science is unambiguous about its claim that the HIV virus causes AIDS. But it has no conception as to why large numbers of individuals that have been infected with the virus for decades do not express the disease? More baffling is the reality of terminal cancer patients who have recovered their lives through spontaneous remissions. Because such remissions are outside the bounds of conventional theory, science completely disregards the fact that they ever happened. Spontaneous remissions are dismissed as unexplainable exceptions to our current truths or, simply, misdiagnoses.

When Positive Thinking Goes Bad

Before I go on to discuss the incredible pow
how my research on cells provided insight in...
mind-body pathways work, I need to make it very clear th...
not believe that simply thinking positive thoughts always leads to
physical cures. You need more than just "positive thinking" to harness control of your body and your life. It *is* important for our health
and well-being to shift our mind's energy toward positive, life-
generating thoughts and eliminate ever-present, energy-draining,
and debilitating negative thoughts. But, and I mean that in the biggest sense of "BUT," the mere thinking of positive thoughts will not
necessarily have any impact on our lives at all! In fact, sometimes
people who "flunk" positive thinking become *more* debilitated
because now they think their situation is hopeless—they believe
they have exhausted all mind and body remedies.

What those positive-thinking dropouts haven't understood
is that the seemingly "separate" subdivisions of the mind, the
conscious and the *subconscious,* are interdependent. The conscious
mind—which represents the seat of our personal identity, source,
or spirit—is the creative mind. It can see into the future, review the
past, or disconnect from the present moment as it solves problems
in our head. In its creative capacity, the conscious mind holds our
wishes, desires, and aspirations for our lives. It is the mind that
conjures up our "positive thoughts."

In contrast, the subconscious mind is primarily a repository of
stimulus-response tapes derived from instincts and learned experiences. The subconscious mind is fundamentally habitual; it will
play the same behavioral responses to life's signals over and over
again, much to our chagrin. How many times have you found yourself going ballistic over something trivial like an open toothpaste
tube? You have been trained since childhood to carefully replace
the cap. When you find the tube with its cap left off, your "buttons are pushed" and you automatically fly into a rage. You've just
experienced the simple stimulus-response of a behavior program
stored in the subconscious mind.

When it comes to sheer neurological processing abilities, the subconscious mind is more than a million times more powerful than the conscious mind. If the desires of the conscious mind conflict with the programs in the subconscious mind, which "mind" do you think will win out? You can repeat the positive affirmation that you are lovable over and over or that your cancer tumor will shrink. But if, as a child, you repeatedly heard that you were worthless and sickly, those messages programmed in your subconscious mind will undermine your best conscious efforts to change your life. Remember how quickly your last New Year's resolution to eat less food fell by the wayside as the aroma of the baking turkey dissolved your resolve?

I believe the greatest problem we face is that we think we are running our lives with the wishes, desires, and aspirations created by our conscious mind. When we struggle or fail to obtain our goals, we are generally led to conclude that we are victims of outside forces preventing us from reaching our destination. However, neuroscience has now established that the conscious mind runs the show, at best, only about 5 percent of the time. It turns out that the programs acquired by the subconscious mind shape 95 percent or more of our life experiences. (Szegedy-Maszak 2005)

Since subconscious programs operate without the necessity of observation or control by the conscious mind, we are completely unaware that our subconscious minds are making our everyday decisions. Our lives are essentially a printout of our subconscious programs, behaviors that were fundamentally acquired from others (our parents, family, and community) before we were six years old. As psychologists recognize, a majority of these developmental programs are limiting and disempowering.

We'll learn more about the origins of self-sabotaging subconscious programming in Chapter 7, "Conscious Parenting." But for the moment, be aware that there is hope even for those of you who used positive thinking and failed miserably. The profoundly important fact is that disempowering programs in the subconscious mind can be quickly rewritten using techniques such as hypnotherapy, affirmations, body-centered therapies, and a large number of new modalities collectively referred to as "energy psychology." I provide a website listing many reprogramming resources in the Addendum section of this book.

Mind Over Body

Let's review what we know about cells. We learned in earlier chapters that the functions of cells are directly derived from the movements of their protein "gears." The movement generated by assemblies of proteins provides the physiologic functions that enable life. While proteins are the physical building blocks, complementary environmental signals are required to animate their movement. The interface between environmental signals and behavior-producing cytoplasmic proteins is the cell's membrane. The membrane receives stimuli and then engages the appropriate, life-sustaining cellular responses. The cell membrane operates as the cell's "brain." Integral membrane receptor-effector proteins (IMPs) are the fundamental physical subunits of the cellular brain's "intelligence" mechanism. By functional definition, these protein complexes are "perception switches" that link reception of environmental stimuli to response-generating protein pathways.

Cells generally respond to an assortment of very basic "perceptions" of what's going on in their world. Such perceptions include whether things like potassium, calcium, oxygen, glucose, histamine, estrogen, toxins, light, or any number of other stimuli are present in their immediate environment. The simultaneous interactions of tens of thousands of reflexive perception switches in the membrane, each directly reading an individual environmental signal, collectively create the complex behavior of a living cell.

For the first three billion years of life on this planet, the biosphere consisted of free-living single cells such as bacteria, algae, and protozoans. While we have traditionally considered such life forms as solitary individuals, we are now aware that signal molecules used by individual cells to regulate their own physiologic functions, when released into the environment, also influence the behavior of other organisms. Signals released by cells into the environment allow for a coordination of behavior among a dispersed population of unicellular organisms. Secreting signal molecules into the environment enhanced the survival of single cells by providing them with the opportunity to live as a primitive dispersed "community."

The single-celled slime mold amoebas provide an example of how signaling molecules lead to community. These amoebas live a solitary existence in the soil foraging for food. When available food in the environment is consumed, the cells synthesize an excess amount of a metabolic by-product called cyclic AMP (cAMP), much of which is released into the environment. The concentration of the released cAMP builds in the environment as other amoebas face starvation. When secreted cAMP signal molecules bind to cAMP-receptors on the cell membranes of other slime mold amoebas, it signals them to activate a swarming behavior wherein the amoebas congregate and form a large multicellular "slug." The slug community is the reproductive stage of slime mold. During the "famine" period, the community of aging cells shares their DNA and creates the next generation of offspring. The new amoebas hibernate as inactive spores. When more food is available, the food molecules act as a signal to break the hibernation, releasing a new population of single cells to start the cycle over again.

The point is that single-celled organisms actually live in a community when they share their "awareness" and coordinate their behaviors by releasing "signal" molecules into the environment. Cyclic AMP was one of evolution's earliest forms of secreted regulatory signals that control cell behavior. The fundamental human signal molecules (e.g., hormones, neuropeptides, cytokines, growth factors) that regulate our own cellular communities were once thought to have arisen with the appearance of complex multicellular life forms. However, recent research has revealed that primitive single-celled organisms were already using these "human" signal molecules in the earliest stages of evolution. (Naokuni and Kanji 1993; Burton, et al, 2002; Kawashima, et al, 2007)

Through evolution, cells maximized the number of IMP "awareness" proteins their membranes could hold. To acquire more awareness, and therefore increase their probability of surviving, cells started to assemble, first into simple colonies and later into highly organized cellular communities. As described earlier, the physiologic functions of multicellular organisms are parceled out to specialized communities of cells forming the body's tissues and organs. In communal organizations, the cell membrane's intelligence processing is carried out by the specialized cells of the organism's nervous and immune systems.

It was only 700 million years ago, recent in regard to the time frame of life on this planet, when single cells found it advantageous to join together in tightly knit multicellular communities, organizations we recognize as animals and plants. The same coordinating signal molecules used by free-living cells were used in these newly evolved closed communities. By tightly regulating the release and distribution of these function-controlling signal molecules, the community of cells would be able to coordinate their functions and act as a single life form. In the more primitive multicellular organisms, those without specialized nervous systems, the flow of these signal molecules within the community provided an elementary "mind," represented by the coordinating information shared by every cell. In such organisms, each cell directly read environmental cues and personally adjusted its own behavior.

However, when cells came together in community, a new politic had to be established. In community, each cell cannot act as an independent agent that does whatever it wants. The term "community" implies that all of its members commit to a common plan of action. In multicellular animals, individual cells may "see" the local environment outside of their own "skin," but they may have no awareness of what is going on in more distant environments, especially those outside of the whole organism itself. Can a liver cell buried in your viscera, responding to its local environmental signals, make an informed response regarding the consequence of a mugger that jumps into your environment? The complex behavior controls needed to ensure a multicellular organization's survival are incorporated within its centralized information processing system.

As more complex animals evolved, specialized cells took over the job of monitoring and organizing the flow of the behavior regulating signal molecules. These cells provided a distributed nerve network and central information processor, a brain. The brain's function is to coordinate the dialogue of signal molecules within the community. Consequently, in a community of cells, each cell must relinquish control to the informed decisions of its awareness authority, the *brain*. The brain *controls* the behavior of the body's cells. This is a very important point to consider as we blame the cells of our organs and tissues for the health issues we experience in our lives.

Emotions: Feeling the Language of Cells

In higher, more aware life forms, the brain developed a specialization that enabled the whole community to tune into the status of its regulatory signals. The evolution of the limbic system provided a unique mechanism that converted the chemical communication signals into sensations that could be experienced by all of the cells in the community. Our conscious mind experiences these *signals* as emotions. The conscious mind not only "reads" the flow of the cellular coordinating signals that comprise the body's "mind"; it can also generate emotions, which are manifested through the controlled release of regulatory signals by the nervous system.

At the same time that I was studying the mechanics of the cell's brain and gaining insight into the operation of the human brain, Candace Pert was studying the human brain and becoming aware of the mechanics of the cell's brain. In *Molecules of Emotion,* Pert revealed how her study of information-processing receptors on nerve cell membranes led her to discover that the same "neural" receptors were present on most, if not all, of the body's cells. Her elegant experiments established that the "mind" was not focused in the head but was distributed via signal molecules to the whole body. As importantly, her work emphasized that emotions were not only derived through a feedback of the body's environmental information. Through self-consciousness, the mind can use the brain to *generate* "molecules of emotion" and override the system. While proper use of consciousness can bring health to an ailing body, inappropriate unconscious control of emotions can easily make a healthy body diseased, a topic I will expand upon in Chapters 6 and 7. *Molecules of Emotion* is a very insightful book describing the scientific discovery process. It also provides some revealing insights into the struggles encountered when trying to introduce new "ideas" into science's Old Boys Club, a subject with which I am all too familiar! (Pert 1997)

The limbic system offered a major evolutionary advance through its ability to sense and coordinate the flow of behavior-regulating signals within the cellular community. As the internal signal system evolved, its greater efficiency enabled the brain to increase in size. Multicellular organisms gained increasingly more cells that were dedicated to responding to an ever-wider variety of *external* environmental signals. While individual cells can respond to simple sensory perceptions such as red, round, aromatic, and sweet, the extra brainpower available in multicellular animals enables them to combine those simple sensations into a higher level of complexity and perceive "apple."

Fundamental reflex behaviors acquired through evolution are passed on to offspring in the form of genetic-based instincts. The evolution of larger brains, with their increased neural cell population, offered organisms the opportunity not only to rely on instinctual behavior, but also to learn from their life experiences. The learning of novel reflex behaviors is essentially a product of *conditioning.* For example, consider the classic example of Pavlov training his dogs to salivate at the ring of a bell. He first trained them by ringing a bell and coupling that stimulus with a food reward. After a while, he would ring the bell but not offer the food. By that time, the dogs were so programmed to expect the food that when the bell rang, they reflexively started to salivate even though no food was present. This is clearly an "unconscious," learned reflex behavior.

Reflex behaviors may be as simple as the spontaneous kick of the leg when a mallet taps the knee or as complex as driving a car at sixty-five miles per hour on a crowded interstate highway while your conscious mind is fully engaged in conversation with a passenger. Though conditioned behavioral responses may be inordinately complex, they are "no-brainers." Through the conditioned learning process, neural pathways between eliciting stimuli and behavioral responses become hardwired to ensure a repetitive pattern. Hardwired pathways are "habits." In lower animals, the entire brain is designed to engage in purely habitual responses to stimuli. Pavlov's dogs salivate by reflex . . . not by deliberate intention. The actions of the subconscious mind are reflexive in nature and are not governed by reason or thinking. Physically, this mind is associated with the activities of *all* of the brain structures that are present in animals that have not evolved self-consciousness.

Humans and a number of other higher mammals have evolved a specialized region of the brain associated with thinking, planning, and decision-making called the prefrontal cortex. This portion of the forebrain is apparently the seat of the "self-conscious" mind processing. The self-conscious mind is self-reflective; it is a newly evolved "sense organ" that observes our own behaviors and emotions. The self-conscious mind also has access to most of the data stored in our long-term memory bank. This is an extremely important feature allowing our history of life to be considered as we consciously plan our futures.

Endowed with the ability to be self-reflective, the self-conscious mind is extremely powerful. It can observe any programmed behavior we are engaged in, evaluate the behavior, and consciously decide to change the program. We can actively *choose* how to respond to most environmental signals and whether we even want to respond at all. The conscious mind's capacity to override the subconscious mind's preprogrammed behaviors is the foundation of free will.

However, our special gift comes with a special pitfall. While almost all organisms have to actually experience the stimuli of life firsthand, the human brain's ability to "learn" perceptions is so advanced that we can actually acquire perceptions indirectly from teachers. Once we accept the perceptions of others as "truths," *their* perceptions become hardwired into our own brains, becoming *our* "truths." Here's where the problem arises: what if our teachers' perceptions are inaccurate? In such cases, our brains are then downloaded with misperceptions. The subconscious mind is strictly a stimulus-response playback device; there is no "ghost" in that part of the "machine" to ponder the long-term consequences of the programs we engage. The subconscious works only in the "now." Consequently, programmed misperceptions in our subconscious mind are not "monitored" and will habitually engage us in inappropriate and limiting behaviors.

If I included as a bonus in this chapter a slithering snake that pops out of this page right now, most of you would run from the room or throw the book out of the house. Whoever "introduced" you to your first snake may have behaved in such a shocking way as to give your impressionable mind an apparently important life lesson: see snake . . . snake baaad! The subconscious memory system is very partial to rapidly downloading and emphasizing perceptions regarding things in your environment that are threatening to life and limb. If you were taught that snakes are dangerous, any time a snake comes into your proximity, you reflexively (unconsciously) engage in a protective response.

But what if a herpetologist were reading this book and a snake popped out? No doubt herpetologists would not only be intrigued by the snake, they would be *thrilled* with the bonus included in the book. Or at least they'd be thrilled once they figured out that the book's snake was harmless. They would then hold it and watch its behaviors with delight. They would think that *your* programmed response was an irrational one because not all snakes are dangerous. Further they would be saddened by the fact that so many people are deprived of the pleasure of studying such interesting creatures. Same snake, same stimulus, yet greatly different responses.

Our responses to environmental stimuli are indeed controlled by perceptions, but not all of our learned perceptions are accurate. Not all snakes are dangerous! Yes, perception "controls" biology, but as we've seen, these perceptions can be true or false. Therefore, we would be more accurate to refer to these controlling perceptions as *beliefs*.

Beliefs control biology!

Ponder the significance of this information. We have the capacity to consciously evaluate our responses to environmental stimuli and change old responses any time we desire . . . once we deal with the powerful subconscious mind, which I discuss in more depth in Chapter 7. We are not stuck with our genes or our self-defeating behaviors!

How the Mind Controls the Body

My insights into how beliefs control biology are grounded in my studies of cloned endothelial cells, the cells that line the blood vessels. The endothelial cells I grew in culture monitor their world closely and change their behavior based on information they pick up from the environment. When I provided nutrients, the cells would gravitate toward those nutrients with the cellular equivalent of open arms. When I created a toxic environment, the cultured cells would retreat from the stimulus in an effort to wall themselves off from the noxious agents. My research focused on the membrane perception switches that controlled the shift from one behavior to the other.

The primary switch I was studying has a protein receptor that responds to histamine, a molecule that the body uses in a way that is equivalent to a local emergency alarm. I found that there are two varieties of switches, H1 and H2, that respond to the same histamine signal. When activated, switches with H1 histamine receptors evoke a *protection response,* the type of behavior revealed by cells in toxin-containing culture dishes. Switches containing H2 histamine receptors evoke a *growth response* to histamine, similar to the behavior of cells cultured in the presence of nutrients.

I subsequently learned that the body's system-wide emergency response signal, adrenaline, also has switches sporting two different adrenaline-sensing receptors, called *alpha* and *beta.* The adrenaline receptors provoked the exact same cell behaviors as those elicited by histamine. When the adrenal *alpha*-receptor is part of an IMP switch, it provokes a protection response when adrenaline is perceived. When the *beta*-receptor is part of the switch, the same adrenaline signal activates a growth response. (Lipton, et al, 1992)

All that was interesting, but the most exciting finding was when I simultaneously introduced both histamine and adrenaline into my tissue cultures. I found that adrenaline signals, released by the central nervous system, override the influence of histamine signals that are produced locally. This is where the politics of the community described earlier come in to play. Suppose you're working in a bank. The branch manager gives you an order. The CEO walks in and gives you the opposite order. Which order would you follow?

If you want to keep your job you'll snap to the CEO's order. There is a similar priority built into our biology, which requires cells to follow instructions from the head honcho nervous system, even if those signals are in conflict with local stimuli.

I was excited by my experiments because I believed that they revealed on the single-cell level a truth for multicellular organisms—that the mind (acting via the central nervous system's adrenaline) overrides the body (acting via the local histamine signal). I wanted to spell out the implications of my experiments in my research paper, but my colleagues almost died from apoplexy at the notion of injecting the body-mind connection into a paper about cell biology. So I put in a cryptic comment about understanding the significance of the study, but I couldn't say what the significance was. My colleagues did not want me to include these implications of my research because the mind is not an acceptable biological concept. The majority of bioscientists are conventional Newtonians—if it isn't matter, it doesn't count. The "mind" is a nonlocalized energy and therefore is not relevant to materialistic biology. Unfortunately, that bias is a "belief" that has been proven to be patently incorrect in a quantum mechanical universe!

Placebos: The Belief Effect

Every medical student learns, at least in passing, that the mind can affect the body. They learn that some people get better when they *believe* (falsely) they are getting medicine. When patients get better by ingesting a sugar pill, medicine defines it as the *placebo effect*. My friend Rob Williams, founder of PSYCH-K, an energy-based psychological treatment system, suggests that it would be more appropriate to refer to it as the *perception effect*. I call it the *belief effect* to stress that our perceptions, whether they are accurate or inaccurate, equally impact our behavior and our bodies.

I celebrate the *belief effect,* which is an amazing testament to the healing ability of the body/mind. However, the "all in their minds" placebo effect has been linked by traditional medicine to, at worst, quacks or, at best, weak, suggestible patients. The placebo effect is quickly glossed over in medical schools so that students can get to the real *tools* of modern medicine like drugs and surgery.

This is a giant mistake. The placebo effect should be a major topic of study in medical school. I believe that medical education should train doctors to recognize the power of our internal resources. Doctors should not dismiss the power of the mind as something inferior to the power of chemicals and the scalpel. They should let go of their conviction that the body and its parts are essentially stupid and that we need outside intervention to maintain our health.

The placebo effect should be the subject of major, funded research efforts. If medical researchers could figure out how to leverage the placebo effect, they would hand doctors an efficient, energy-based, side-effect-free tool to treat disease. Energy healers say they already have such tools, but I am a scientist, and I believe the more we know about the science of the placebo, the better we'll be able to use it in clinical settings.

I believe the reason the mind has so summarily been dismissed in medicine is the result not only of dogmatic thinking, but also of financial considerations. If the power of your mind can heal your sick body, why should you go to the doctor and, *more* importantly, why would you need to buy drugs? In fact, I was recently chagrined to learn that drug companies are studying patients who respond to sugar pills with the goal of *eliminating* them from early clinical trials. It inevitably disturbs pharmaceutical manufacturers that in most of their clinical trials the placebos, the "fake" drugs, prove to be as effective as their engineered chemical cocktails. (Greenberg 2003) Though the drug companies insist they're not trying to make it easier for ineffective drugs to get approved, it is clear that effectiveness of placebo pills is a threat to the pharmaceutical industry. The message from the drug companies is clear to me: if you can't beat placebo pills fairly, simply remove the competition!

The fact that most doctors are not trained to consider the impact of the placebo effect is ironic because some historians make a strong case that the history of medicine is largely the history of the placebo effect. For most of medical history, doctors did not have effective methods to fight disease. Some of the more notorious treatments once prescribed by mainstream medicine include blood-letting, treating wounds with arsenic, and the proverbial cure-all, rattlesnake oil. No doubt some patients, the conservatively esti-mated one third of the population who are particularly susceptible to the healing power of the placebo effect, got better with those treatments. In today's world, when doctors wearing white coats deliver a treatment decisively, patients may *believe* the treatment works—and so it does, whether it is a real drug or just a sugar pill.

Though the question of *how* placebos work has in the main been ignored by medicine, recently some mainstream medical researchers are turning their attention to it. (Erdmann 2008; Price, et al, 2008; Niemi 2009) The results of those studies suggest that it is not only wacky, nineteenth-century treatments that can foster a placebo effect but also modern medicine's sophisticated technol-ogy, including the most "concrete" of medical tools, surgery.

A Baylor School of Medicine study, published in 2002 in the *New England Journal of Medicine,* evaluated surgery for patients with severe, debilitating knee pain. (Moseley, et al, 2002) The lead author of the study, Dr. Bruce Moseley, "knew" that knee surgery helped his patients: "All good surgeons know there is no placebo effect in surgery." But Moseley was trying to figure out which part of the surgery was giving his patients relief. The patients in the study were divided into three groups. Moseley shaved the damaged cartilage in the knee of one group. For another group, he flushed out the knee joint, removing material thought to be causing the inflammatory effect. Both of these constitute standard treatment for arthritic knees. The third group got "fake" surgery. The patient was sedated, Moseley made three standard incisions and then talked and acted just as he would have during a real surgery—he even splashed salt water to simulate the sound of the knee-washing procedure. After forty minutes, Moseley sewed up the incisions as if he had done the surgery. All three groups were prescribed the same postoperative care, which included an exercise program.

The results were shocking. Yes, the groups who received sur-gery, as expected, improved. But the placebo group improved just as much as the other two groups! Despite the fact that there are 650,000 surgeries yearly for arthritic knees, at a cost of about $5,000 each, the results were clear to Moseley: "My skill as a surgeon had no benefit on these patients. The entire benefit of surgery for osteoarthritis of the knee was the placebo effect." Television news programs graphically illustrated the stunning results. Footage showed members of the placebo group walking and playing basketball, in short doing things they reported they could not do before their "surgery." The placebo patients didn't find out for two years that they had gotten fake surgery. One member of the placebo group, Tim Perez, who had to walk with a cane before the surgery, is now able to play basketball with his grandchildren. He summed up the theme of this book when he told the Discovery Health Channel: "In this world anything is possible when you put your mind to it. I know that your mind can work miracles."

Studies have shown the placebo effect to be powerful in treating other diseases, including asthma and Parkinson's. In the treatment of depression, placebos are stars. So much so that psychiatrist Walter Brown of the Brown University School of Medicine has proposed placebo pills as the first treatment for patients with mild or moderate depression. (Brown 1998) Patients would be told that they're getting a remedy with no active ingredient, but that shouldn't dampen the pills' effectiveness. Studies suggest that even when people know they're not getting a drug, the placebo pills still work.

One indication of the power of the placebo came from a report from the United States Department of Health and Human Services. The report found that half of severely depressed patients taking drugs improve versus 32 percent taking a placebo. (Horgan 1999) Even that impressive showing may underestimate the power of the placebo effect: many study participants figure out they're taking the real drug because they experience side effects that are not experienced by those taking the placebo. Once those patients realize they're taking the drug, i.e., once they start *believing* that they're getting the *real* pill, they are particularly more susceptible to the placebo effect.

Given the power of the placebo, it is no wonder that the $8.2 billion antidepressant industry is under attack by critics who charge that drug companies are hyping the effectiveness of their pills. In a

2002 article in the American Psychological Association's *Prevention & Treatment,* "The Emperor's New Drugs," University of Connecticut psychology professor Irving Kirsch found that 80 percent of the effect of antidepressants, as measured in clinical trials, could be attributed to the placebo effect. (Kirsch, et al, 2002) Kirsch had to invoke the Freedom of Information Act in 2001 to get information on the clinical trials of the top antidepressants: these data were not forthcoming from the Food and Drug Administration. The data show that in more than half of the clinical trials for the six leading antidepressants, the drugs did not outperform placebo, sugar pills. And Kirsch noted in a Discovery Health Channel interview that "the difference between the response of the drugs and the response of placebo was less than two points on average on this clinical scale that goes from fifty to sixty points. That's a very small difference. That difference clinically is meaningless."

Another interesting fact about the effectiveness of antidepressants is that they have performed better and better in clinical trials over the years, suggesting that their placebo effects are in part due to savvy marketing. The more the miracle of antidepressants was touted in the media and in advertisements, the more effective they became. Beliefs are contagious! We now live in a culture where people *believe* that antidepressants work, and so they do.

A California interior designer, Janis Schonfeld, who took part in a clinical trial to test the efficacy of Effexor in 1997, was just as "stunned" as Perez when she found out that she had been on a placebo. Not only had the pills relieved her of the depression that had plagued her for thirty years, the brain scans she received throughout the study found that the activity of her prefrontal cortex was greatly enhanced. (Leuchter, et al, 2002) Her improvements were not "all in her head." When the mind changes, it absolutely affects your biology. Schonfeld also experienced nausea, a common Effexor side effect. She is typical of patients who improve with placebo treatment and then find out they were not on the real drug—she was convinced the doctors had made a mistake in the labeling for she "knew" she was on the drug. She insisted that the researchers double-check their records to make absolutely sure she wasn't on the drug.

Nocebos: The Power of Negative Beliefs

While many in the medical profession are aware of the placebo effect, few have considered its implications for self-healing. If positive thinking can pull you out of depression and heal a damaged knee, consider what negative thinking can do in your life. When the mind, through positive suggestion, improves health, it is referred to as the placebo effect. Conversely, when the same mind is engaged in negative suggestions that can damage health the negative effects are referred to as the *nocebo* effect.

In medicine, the nocebo effect can be as powerful as the placebo effect, a fact you should keep in mind every time you step into a doctor's office. By their words and their demeanor, physicians can convey hope-deflating messages to their patients, messages that are, I believe, completely unwarranted. Albert Mason, for example, thinks his inability to project optimism to his patients hampered his efforts with his ichthyosis patients. Another example is the potential power of the statement: "You have six months to live." If you choose to believe your doctor's message, you are not likely to have much more time on this Earth.

I have cited the Discovery Health Channel's 2003 program "Placebo: Mind Over Medicine" in this chapter because it is a good compendium of some of medicine's most interesting cases. One of its more poignant segments featured a Nashville physician, Clifton Meador, who has been reflecting on the potential power of the nocebo effect for thirty years. In 1974 Meador had a patient, Sam Londe, a retired shoe salesman suffering from cancer of the esophagus, a condition that was at the time considered 100 percent fatal. Londe was treated for that cancer, but everyone in the medical community "knew" that his esophageal cancer would recur. So it was no surprise when Londe died a few weeks after his diagnosis.

The surprise came after Londe's death when an autopsy found very little cancer in his body, certainly not enough to kill him. There were a couple of spots in the liver and one in the lung, but there was no trace of the esophageal cancer that everyone thought had killed him. Meador told the Discovery Health Channel: "He died with cancer, but not from cancer." What did Londe die of if not esophageal cancer?

Had he died because he *believed* he was going to die? The case still haunts Meador three decades after Londe's death: "I thought he had cancer. He thought he had cancer. Everybody around him thought he had cancer . . . did I remove hope in some way?" Troublesome nocebo cases suggest that physicians, parents, and teachers can remove hope by programming you to believe you are powerless.

Our positive and negative beliefs not only impact our health but also every aspect of our life. Henry Ford was right about the efficiency of assembly lines, and he was right about the power of the mind: "If you believe you can or if you believe you can't . . . you're right." Think about the implications of the man who blithely drank the bacteria that medicine had decided caused cholera. Consider the people who walk across coals without getting burned. If they wobble in the steadfastness of their belief that they can do it, they wind up with burned feet. Your beliefs act like filters on a camera, changing how you see the world. And your biology adapts to those beliefs. When we truly recognize that our beliefs are that powerful, we hold the key to freedom. While we cannot readily change the codes of our genetic blueprints, we can change our minds and, in the process, switch the blueprints used to express our genetic potential.

In my lectures I provide two sets of plastic filters, one red and one green. I have the audience pick one color and then look at a blank screen. I then tell them to yell out whether the image I project next is one that generates love or generates fear. Those in the audience that don the red "belief" filters see an inviting picture of a cottage labeled "House of Love," flowers, a sunny sky and the message: "I live in Love." Those wearing the green filters see a threatening dark sky, bats, snakes, a ghost hovering outside a dark, gloomy house, and the words: "I live in fear." I always get enjoyment out of seeing how the audience responds to the confusion when half yell out: "I live in love," and the other half, in equal certainty, yells out: "I live in fear" in response to the same image.

Then I ask the audience to change to the opposite colored filters. My point is that you can choose what to see. You can filter your life with rose-colored beliefs that will help your body grow or you can use a dark filter that turns everything black and makes your body/mind more susceptible to disease. You can live a life of fear or live a life of love. You have the choice! But I can tell you that if you choose to see a world full of love, your body will respond by growing in health. If you choose to believe that you live in a dark world full of fear, your body's health will be compromised as you physiologically close yourself down in a protection response.

Learning how to harness your mind to promote growth is the secret of life, which is why I called this book *The Biology of Belief.* Of course the secret of life is not a secret at all. Teachers like Buddha and Jesus have been telling us the same story for millennia. Now science is pointing in the same direction. It is not our genes but our beliefs that control our lives . . . Oh ye of little belief!

That thought is a good entrée into the next chapter, in which I'll detail how living in love and living in fear create opposite effects in the body and the mind. Before we leave this chapter, I'd just like to emphasize again that there is nothing wrong with going through life wearing the proverbial rose-colored glasses. In fact, those rose-colored glasses are necessary for your cells to thrive. Positive thoughts are a biological mandate for a happy, healthy life. In the words of Mahatma Gandhi:

Your beliefs become your thoughts
Your thoughts become your words
Your words become your actions
Your actions become your habits
Your habits become your values
Your values become your destiny

✳ ✳ ✳

Since the publication of *The Biology of Belief's* first edition, a whole new field of research called behavioral epigenetics has emerged that is unraveling the mechanisms that explain how donning rose-colored glasses and fostering social connections can enable your cells to thrive. The mission of behavioral epigenetic scientists is nothing less than to figure out how nurture shapes nature. Here, nature refers to gene-controlled characteristics, and nurture refers to the influence of a wide range of life experiences, from social interactions to nutrition to a positive mental attitude.

This research has confirmed that brain cells translate the mind's perceptions (beliefs) of the world into complementary and unique chemical profiles that, when secreted into the blood, control the fate of the body's 50 trillion cells. So blood, the body's culture medium, not only nourishes cells, its neurochemical components also regulate cells' genetic and behavioral activity. As Steve Cole, an epigeneticist at UCLA's School of Medicine, told *Pacific Standard* magazine: "A cell is a machine for turning experience into biology." (Dobbs 2013)

When we change the way we perceive the world, that is, when we "change our beliefs," we change the blood's neurochemical composition, which then initiates a complementary change in the body's cells. *The function of the mind is to create coherence between our beliefs and the reality we experience.* That explains why my health and energy soared after I jettisoned my old, "depressed, fatalistic" view of the world that I talked about in the Prologue.

Despite medicine's resistance to conceding the crucial role our minds play in our health, science has long had to face the fact that some physiologic systems, primarily the body's skeletal musculature, are actually under voluntary control of the conscious mind. And there have always been phenomena that called into question the biomedical belief that the remaining bodily functions are under the *involuntary* control of the autonomic nervous system. When yogis demonstrated that they could consciously override autonomic controls, such as the regulation of body temperature, blood pressure, and pH, they provided evidence of the conscious mind's ability to influence the body's innate intelligence. So did hypnotists when they told individuals in a trance that a burning cigarette touched them. Though in reality they were touched only with the hypnotists' fingertips, the individuals expressed a full burn response in the form of a blister and wheal and flare (inflammation of the skin). (Paul 1963) Clearly an individual's belief, in this case the *misperception* of being burned, resulted in a complex and formerly perceived autonomic burn response in healthy skin, just as Dr. Albert Mason's misperceptions that opened this chapter cured an incurable disease.

The conclusion is simple: positive perceptions of the mind enhance health by engaging immune functions, while inhibition of immune activities by negative perceptions can precipitate dis-ease. Those negative perceptions can also create debilitating, chronic psychological stress that has a profound and negative impact on gene function. Research on mice has shown, for example, that long-term exposure to stress hormones leaves a lasting mark on the genome and modulates the behavior of genes that control mood and behavior. To see if stress might epigenetically influence genes involved in depression, the drinking water of one group of mice was spiked with corticosterone (the rodent version of cortisol) for four weeks. Control mice drank plain water without this glucocorticoid hormone. Mice who received corticosterone displayed characteristics of anxiety in behavioral tests. Assessment of gene activity showed that these mice had a significant increase in *Fkbp5*, a protein whose human equivalent has been linked to mood disorders, including depression and bipolar disease. (Lee 2010)

Our stress responses were designed for intermittent use such as escaping from the occasional saber-toothed tiger. The chronic nature of modern stress that occurs 24-7-365 epigenetically taxes our stress response mechanisms and leads to depression or other mood disorders. Unsurprisingly, Dr. Herbert Benson, famed Mind/Body Medical Institute Professor of Medicine at Harvard Medical School, has concluded that stress is responsible for up to 90 percent of all doctor office visits. (Benson 1997)

Though stress plays a major role as a risk factor in disease, UCLA epigeneticist Cole, who was one of the earliest researchers to bring the study of whole genome expression into the realm of social psychology, has concluded that social isolation is an even more potent and underestimated risk factor. "If you actually measure stress, using our best available instruments, it can't hold a candle to social isolation. Social isolation is the best-established, most robust social or psychological risk factor for disease out there. Nothing can compete," he told *Pacific Standard* magazine. (Dobbs 2013)

Cole has discovered that whole sectors of genes look markedly different in lonely people versus people who are socially secure. Of the approximately 19,000 human genes, lonely and not-lonely people showed sharply different gene expression responses in 209 genes, many of which play roles in inflammatory immune responses. He reasoned that if social stress reliably created this immune gene profile, it might explain the results of his earlier studies in which lonely HIV carriers succumbed so much faster to the disease than socially active HIV carriers. (Cole, et al, 1996)

Cole further discovered the destructive influence of social stress when he and collaborators interviewed 103 healthy Vancouver-area women aged fifteen to nineteen, recorded information about chronic interpersonal stress on their lives, drew blood, and ran gene expression profiles. Six months later, blood was drawn and the gene profiles run again. From the results of their first social stress level tests, researchers were able to predict which women would show changes in their gene activity when measured six months later. The findings of Cole's team suggest "that chronic interpersonal difficulties accentuate expression of pro- and anti-inflammatory signaling molecules" and that "these dynamics may underlie the excess morbidity associated with social stress, particularly in inflammation-sensitive diseases like depression and atherosclerosis." (Miller, et al, 2009)

In another study, Cole and collaborators found a similarly unbalanced gene expression and immune response profile in groups of impoverished children and depressed patients with asthma. The team discovered that immune functions in poorer kids had more active inflammatory genes and, simultaneously, expressed more sluggishness in gene networks that control the inflammation response than well-to-do children. The health histories of the poor kids also showed more asthma attacks and other health problems.

Though poverty seemed to be interfering with the behavior of their immune systems, Cole's team suspected other factors at work. So they showed all the kids films of ambiguous or awkward social situations and asked them how threatening they found them. On average, the poor kids perceived more threat; the well-off children perceived less. But some kids in both groups were outliers: a few of the poor kids saw very little menace in the ambiguous situations and a few well-off kids saw a lot. When the results of individuals with perceptions of insecurity were separated from the participants' socioeconomic scores and laid over the gene-expression scores, the data showed that it was really the kids' *perceptions* of their vulnerability, their *perceptions* of how scary the world is, not their income levels, that accounted for most of the difference in immune gene expression. In fact, when controlled for variations in threat perception, poverty's influence almost vanished. (Cole 2009)

The question of why the kids found the world so scary was outside the subject of that study, but Cole believes that isolation plays a key role, a hypothesis buttressed by a 2004 study in which Yale psychiatrist Joan Kaufman studied fifty-seven school-age children who had been removed from their homes because they had been abused. The study measured the serotonin transporter gene (SERT), which has both a long and short form, because previous studies had found that people who carry the short SERT are more likely to become depressed or anxious when stressed. The kids with the short SERT did in fact suffer twice as many mental health problems as those with the long variety. But there were unexpected results as well. When Kaufman laid both the kids' depression scores and their SERT variants across their levels of social support (defined narrowly as contact at least once monthly with a trusted adult figure outside the house), that seemingly paltry (i.e., once a month) social connection erased about 80 percent of the combined risk of the depression-related short SERT variant. (Kaufman, et al, 2004) Science writer David Dobbs asks: "If social connection can almost completely protect us against the well-known effects of serious abuse, isn't the isolation almost as toxic as the beatings and neglect?" (Dobbs 2013)

Many traits mediated by behavioral epigenetics have been shown to carry over to subsequent generations. One example is the long-term health problems that plague many people raised in lower socioeconomic environments, including the vicious cycle in which abused children grow up to be abusive parents and the self-destructive behavior that leads to drug addiction. The implication of behavioral epigenetics research is not that people are doomed to lead dysfunctional lives because their parents did—epigenetic traits are not immutably coded genetic traits.

So the message of this anniversary edition is the same message as the first edition of *The Biology of Belief* and other books that have appeared in its wake—your genes do not dictate your life and you can change your life when you change your beliefs. As acclaimed Harvard lecturer and best-selling author Shawn Achor writes in *The Happiness Advantage: The Seven Principles of Positive Psychology That Fuel Success and Performance at Work,* "The belief that we are just our genes is one of the most pernicious myths in modern culture—the insidious notion that people come into the world with a fixed set of abilities and that they, and their brains, cannot change. The scientific community is partly to blame for this because for decades scientists refused to see what potential for change was staring them right in the face." (Achor 2010)

Achor's studies emphasize that we have been culturally programmed to believe that if we reach our goals (when we get into Harvard, lose twenty pounds, get a high-paying job, etc.), *then* we'll be happy. However, this formula for happiness is actually inverted: happiness *fuels* success, not the other way around. Simply stated, success doesn't bring happiness; happiness brings success. This fact is supported by a wealth of research in psychology and neuroscience that demonstrates that a positive outlook enhances brain activity and leads to a more creative, motivated, and productive work experience.

Achor's studies of 1,600 Harvard students, one in five of whom flourish in Harvard's pressure cooker environment, also support Cole's conclusion about the importance of social connections. Achor tells the stories of two spirited roommates, Amanda and Brittney. Both started freshman year with new friends they made easily, but as midterms approach, they parted ways. Amanda gravitated to a secluded cubicle in the library, isolating herself from her peers. Brittney, on the other hand, organized study groups that included time for small talk, and when she studied alone, she made sure to take study breaks, including a ten-minute break to participate in an Oreo-eating contest. By January, Amanda was wishing she could transfer to a less competitive school and Brittney, who had made a point of keeping up her social connections, was "happy, well-adjusted, and performing exceptionally in her courses." (Achor 2010)

Now for the "good news"—and not just for the small number of people in the world who study at Harvard! Belief modification can induce rapid changes in gene activity. When individuals raise their levels of optimism and deepen their social connections (*à la* Steve Cole and

Brittney), they not only raise their level of happiness, but also dramatically improve every single business and educational outcome tested for.

A recent study revealed that just eight hours of mindful meditation was sufficient to significantly change vital gene functions. Compared to controls, meditators exhibited a range of genetic and molecular differences that included reduced levels of pro-inflammatory genes and altered levels of gene-regulating machinery. These observed changes in genetic expression are associated with faster physical recovery from stressful situations and prove that mindfulness practice can lead to health improvement through profound epigenetic alterations of the genome. (Kaliman, et al, 2014)

Research suggests that a positive mindset can even trump some of the effects of aging. A number of studies have found that people who hold more positive stereotypes about aging behave differently as they age from those who hold more negative stereotypes, even when they are similar in other ways, including how healthy they are. Recently, Yale University and University of California, Berkeley researchers decided to study whether it is possible to counteract our culture's undeniably negative view of aging and, in so doing, improve health. One hundred individuals ranging in age from sixty to ninety-nine years were assigned to one of four groups: (1) a group that experienced implicit positive aging stereotypes intervention, (2) a group that experienced explicit positive aging intervention, (3) a group that experienced both implicit and explicit positive aging stereotype intervention, or (4) a control group. (Levy, et al, 2014)

The stars of the study turned out to be the implicit positive age stereotype intervention group. That group, in four fifteen-minute weekly sessions, unconsciously observed positive words connected to aging. Words like "wise," "creative," "spry," and "fit" connected to "old" and "senior" were flashed on a laptop screen so briefly that while the subconscious mind registered them, the participant's slower-operating conscious mind couldn't perceive the words. Follow-up testing showed that this implicit intervention significantly strengthened positive age stereotypes and self-perceptions of age, but even more impressive were physical changes. One week and three weeks after the final session, participants were given physical tasks: repeatedly standing up from a chair and sitting down, walking across a room, holding poses that challenge balance. The group that had experienced implicit positive messages showed significant improvement in their physical functioning compared to before the experiment began. In fact, they showed greater physical improvement than a group of similar-aged adults who exercised for four months.

Those who participated in the explicit intervention that consisted of writing essays about positive, fit, energetic elders showed no improvement. The results led Yale researcher Becca Levy to speculate that implicit messaging is an effective way to break through this culture's ubiquitous negative stereotypes about aging, beliefs that children "as young as three or four" have already absorbed. People encounter negative stereotypes through media and marketing and everyday conversations so often that they build up ways to hold on to them. Implicit interventions can bypass that. That's because subliminal messages reach the subconscious mind where negative programming about getting old is lodged alongside all the other negative messages absorbed in our youth.

The fact that implicit rather than explicit learning worked emphasizes a vitally important point that I stressed in the original edition of this chapter: the fact that the primary source controlling our life experiences is the *subconscious* mind, and we need to focus on reprogramming it rather than just shifting our conscious mind's beliefs. As a septuagenarian who is not yet ready to get out of the game, I reprogram myself daily lest any belief that aging is compromising my physical or mental activities limit them. While a mirror reflects my aging body, the consciousness behind my eyes does not own that reality and operates from a more ageless point of view. I support my plan of action by simply avoiding mirrors!

Current research echoes another amazing mind-over-genes experiment about aging, conducted more than thirty years ago, of eight men in their seventies who were dropped off at the front entrance of a converted monastery in New Hampshire. A few of them were stooped with arthritis, and two walked with canes. When they entered the building they walked into a time warp. Music on a vintage radio played tunes from 1959, black-and-white programs on the TV showed archived videos of old programs such as *The Ed Sullivan Show*, books on the shelves and scattered magazines lying around were all from the same period. This became the men's home for a five-day, radical experiment designed by Harvard psychologist Ellen Langer. During their stay, the test subjects reminisced and engaged in conversations about events and sports of that time period. (Langer 2009)

Measurement of the men's dexterity, grip strength, flexibility, hearing and vision, memory, and cognition, all testable biomarkers of age, were assessed before they arrived and the men were retested at the end of their stay. By several measurements, they outperformed a control group that came to the monastery but did not participate in the time warp experiment. The experimental group was more supple, possessed greater manual dexterity, and sat taller. Most unexpectedly, their sight improved and independent judges acknowledged that they looked younger. Langer remarked that the men had "put their mind in an earlier time," and their bodies went along for the ride. Unfortunately, the experiment could not be repeated because of its complications and expense.

However, in 2010, the BBC recreated Langer's experiment in a four-part broadcast called "The Young Ones," this time engaging six aging former celebrities as the test subjects. These men were transported in vintage cars to a country house meticulously retrofitted to represent a 1975 home. After a week of reliving and sharing thirty-five-year-old news and sports stories, the aging celebrities showed the same marked improvement on test assessments as the rejuvenated septuagenarian participants in Langer's New Hampshire experiment. One of the test subjects who had arrived in a wheelchair walked out with a cane. Another individual who could not put his socks on without assistance when he arrived hosted the final evening's dinner party, easily moving around with enthusiasm and purpose. Those who stooped when they first arrived left walking taller and looked younger. (Grierson 2014)

The production was nominated for a British Emmy and renewed interest in Langer's research, which is currently being expanded through a variety of approaches, all of which are measuring how a change in the perceptions of time can lead to physiologic and mental "youthing." Psychologist Jeffrey Rediger, a Harvard colleague of Ellen Langer, acknowledged, "health and illness are much more rooted in our minds and in our hearts and how we experience ourselves in the world than our models even begin to understand."

In another mind-bending study in 2007, this one about weight loss, researchers told half of the cleaning staff at seven hotels that they were burning lots of calories in their daily work, enough to satisfy the Surgeon General's recommendations for an active lifestyle; the other half did not hear that positive news. "Although actual behavior did not change, four weeks after the intervention, the informed group *perceived* themselves to be getting significantly more exercise than before." And in fact, those who perceived that they were getting more exercise lost weight and lowered their blood pressure, body fat, waist-to-hip ratio, and body mass index. (Crum and Langer 2007)

The evidence that belief exerts a powerful influence over physiology, gene expression, and behavior has led epigeneticist Cole to conclude: "To an extent that immunologists and psychologists rarely appreciate, we are architects of our own experience. Your subjective experience carries more power than your objective situation." (Dobbs 2013) In Cole's quote, the term "subjective experience" represents *perception* or *belief*, while "objective situation" can be interpreted as *reality*. Replacing Cole's words with these synonyms, his quote now reads: Your *belief* carries more power than your *reality*. Hence . . . *The Biology of Belief!*

CHAPTER 6

GROWTH and PROTECTION

Evolution has provided us with lots of survival mechanisms. They can be roughly divided into two functional categories: growth and protection. These growth and protection mechanisms are the fundamental behaviors required for an organism to survive. I'm sure you know how important it is to protect yourself. You may not realize though that growth is vitally important for your survival as well—even if you're an adult who has reached your full height. Every day billions of cells in your body wear out and need to be replaced. For example, the entire cellular lining of your gut is replaced every seventy-two hours. In order to maintain this continuous turnover of cells, your body needs to expend a significant amount of energy daily.

By now you won't be surprised to learn that I first became aware of how important growth and protection behaviors are in the laboratory where my observations of single cells have so often led me to insights about the multicellular human body. When I was cloning human endothelial cells, they *retreated* from toxins that I introduced into the culture dish, just as humans retreat from mountain lions and muggers in dark alleys. They also *gravitated* to nutrients, just as humans gravitate to breakfast, lunch, dinner, and love. These opposing movements define the two basic cellular responses to environmental stimuli. Gravitating *to* a life-sustaining signal, such as nutrients, characterizes a growth response; moving *away* from threatening signals, such as toxins, characterizes a protection response. It must also be noted that some environmental stimuli are neutral; they provoke neither a growth nor a protection response.

My research at Stanford showed that these growth/protection behaviors are also essential for the survival of multicellular organisms such as humans. But there is a catch to these opposing survival mechanisms that have evolved over billions of years. It turns out that the mechanisms that support growth and protection cannot operate optimally at the same time. In other words, cells cannot simultaneously move forward and backward. The human blood vessel cells I studied at Stanford exhibited one microscopic anatomy for providing nutrition and a completely different microscopic anatomy for providing a protection response. What they couldn't do was exhibit both configurations at the same time. (Lipton, et al, 1991)

In a response similar to that displayed by cells, humans unavoidably restrict their growth behaviors when they shift into a protective mode. If you're running from a mountain lion, it's not a good idea to expend energy on growth. In order to survive—that is, escape the lion—you summon all your energy for your fight-or-flight response. Redistributing energy reserves to fuel the protection response inevitably results in a curtailment of growth.

In addition to diverting energy to support the tissues and organs needed for the protection response, there is an additional reason why growth is inhibited. Growth processes require an open exchange between an organism and its environment. For example, food is taken in and waste products are excreted. However, protection requires a closing down of the system to wall the organism off from the perceived threat.

Inhibiting growth processes is also debilitating in that growth is a process that not only expends energy but is also required to *produce* energy. Consequently, a sustained protection response *inhibits the creation of life-sustaining energy*. The longer you stay in protection, the more you consume your energy reserves, which in turn, compromises your growth. In fact, you can shut down growth processes so completely that it becomes a truism that you can be "scared to death."

Thankfully, most of us don't get to the "scared to death" point. Unlike single cells, the growth/protection response in multicellular organisms is not an either/or proposition—not all of our 50 trillion cells have to be in growth or protection mode at the same time. The proportion of cells in a protection response depends on the severity of the perceived threats. You can survive while under stress from these threats, but chronic inhibition of growth mechanisms severely compromises your vitality. It is also important to note that to fully experience your vitality it takes more than just getting rid of life's stressors. In a growth-protection continuum, eliminating the stressors only puts you at the neutral point in the range. To fully thrive, we must not only eliminate the stressors but also actively seek joyful, loving, fulfilling lives that stimulate growth processes.

The Biology of Homeland Defense

In multicellular organisms, growth/protection behaviors are controlled by the nervous system. It is the nervous system's job to monitor environmental signals, interpret them, and organize appropriate behavioral responses. In a multicellular community, the nervous system acts like the government in organizing the activities of its cellular citizens. When the nervous system recognizes a threatening environmental stress, it alerts the community of cells to impending danger.

The body is actually endowed with two separate protection systems, each vital to the maintenance of life. The first is the system that mobilizes protection against *external* threats. It is called the HPA axis, which stands for the Hypothalamus-Pituitary-Adrenal axis. When there are no threats, the HPA axis is inactive and growth flourishes. However, when the brain's hypothalamus perceives an environmental threat, it engages the HPA axis by sending a signal to the pituitary gland, the "Master Gland," which is responsible for organizing the 50 trillion cells of the community to deal with the impending threat.

Think back to the cell membrane's stimulus-response mechanism, the receptor-effector proteins—the hypothalamus and pituitary gland are behavioral equivalents. Similar to the role of a receptor protein, the hypothalamus receives and recognizes environmental signals; the pituitary's function resembles that of the effector protein in that it launches the body's organs into action. In response to threats from the external environment, the pituitary gland sends a signal to the adrenal glands, informing them of the need to coordinate the body's fight-or-flight response.

The technical details of how stress stimuli engage the HPA axis follow a simple cascade: In response to perceptions of stress registered in the brain, the hypothalamus secretes a corticotropin-releasing factor (CRF), which travels to the pituitary gland. CRF activates special pituitary hormone-secreting cells, causing them to release adrenocorticotropic hormones (ACTH) into the blood. The ACTH then makes its way to the adrenal glands, where it serves as the signal to turn on the secretion of the "fight-flight" adrenal hormones. These stress hormones coordinate the function of the body's organs, providing us with great physiologic power to fend off or flee from danger.

Once the adrenal alarm is sounded, the stress hormones released into the blood constrict the blood vessels of the digestive tract, forcing the energy-providing blood to preferentially nourish the tissues of the arms and legs that enable us to get out of harm's way. Before the blood was sent to the extremities, it was concentrated in the visceral organs. Redistributing the viscera's blood to the limbs in the fight-or-flight response results in an inhibition of growth-related functions; without the blood's nourishment the visceral organs cannot function properly. The visceral organs stop doing their life-sustaining work of digestion, absorption, excretion, and other functions that provide for the growth of the cells and the production of the body's energy reserves. Hence, the stress response inhibits growth processes and further compromises the body's survival by interfering with the generation of vital energy reserves.

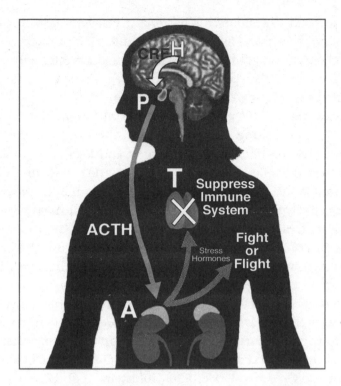

The body's second protection system is the immune system, which protects us from threats originating under the skin, such as those caused by bacteria and viruses. When the immune system is mobilized, it can consume much of the body's energy supply. To get a sense of how much energy the immune system expends, recall how physically weak you become when you are fighting infections such as a flu or a cold. When the HPA axis mobilizes the body for fight-or-flight response, the adrenal hormones directly repress the action of the immune system to conserve energy reserves. In fact, stress hormones are so effective at curtailing immune system function that doctors provide them to recipients of transplants so that their immune systems won't reject the foreign tissues.

Why would the adrenal system shut down the immune system? Imagine that you are in your tent on the African savannah suffering from a bacterial infection and experiencing a bad case of diarrhea. You hear the gutty growl of a lion outside your tent. The brain must make a decision about which is the greater threat. It will do your body no good to conquer the bacteria if you let a lion maul you. So your body halts the fight against the infection in favor of mobilizing energy for flight to survive your close encounter with a lion. Therefore, a secondary consequence of engaging the HPA axis is that it interferes with our ability to fight disease.

Activating the HPA axis also interferes with our ability to think clearly. The processing of information in the forebrain (conscious mind), the center of executive reasoning and logic, is significantly slower than the reflex activity controlled by the hindbrain (subconscious mind). In an emergency, the faster the information processing, the more likely the organism will survive. Adrenal stress hormones constrict the blood vessels in the forebrain, reducing its ability to engage in conscious volitional action. Constriction of forebrain blood vessels redirects vascular flow to the hindbrain. The increase in nutrition and energy enhances the hindbrain's life-sustaining reflexes to more effectively control fight-or-flight behavior. While it is necessary that stress signals repress the slower processing conscious mind to augment survival, it comes at a cost . . . diminished conscious awareness and reduced intelligence. (Takamatsu, et al, 2003; Arnsten and Goldman-Rakic 1998; Goldstein, et al, 1996)

Fear Kills

Remember the shell-shocked, frozen look on my Caribbean medical students' faces when they failed my test, the medical school equivalent of a voracious lion? Had my students stayed frozen in fear, I can guarantee you that they would have performed dismally on their finals. The simple truth is, when you're frightened, you're dumber. Teachers see it all the time among students who "don't test well." Exam stress paralyzes these students who, with trembling hands, mark wrong answers because in their panic, they can't access cerebrally stored information they have carefully acquired all semester.

The HPA system is a brilliant mechanism for handling acute stresses. However, this protection system was not designed to be continuously activated. In today's world, most of the stresses we are experiencing are not in the form of acute, concrete "threats" that we can easily identify, respond to, and move on. We are constantly besieged by multitudes of unresolvable worries about our personal lives, our jobs, and our war-torn global community. Such worries do not threaten our immediate survival, but they nevertheless can activate the HPA axis, resulting in chronically elevated stress hormones.

To illustrate the adverse effects of sustained adrenaline, let's use an example of a track race. An extremely well-trained and healthy group of sprinters step up to the starting line. When they hear the command "On your mark!" they get on their hands and knees and adjust their feet into the starting blocks. Then the starter barks out, "Get set." The athletes' muscles tighten as they prop themselves up on their fingers and toes. When they shift into "Get set" mode, their bodies release the flight-promoting adrenaline hormones that power their muscles for the arduous task ahead. While the athletes are on hold awaiting the "Go" command, their bodies are straining in anticipation of that task. In a normal race, that strain lasts only a second or two before the starter yells, "Go!" However, in our mythical race, the "Go" command, which would launch the athletes into action, never comes. The athletes are left in the starting blocks, their blood coursing with adrenaline, their bodies fatiguing with the strain of preparing for the race that never comes. No matter how toned their physique, within seconds, these athletes will physically collapse from the strain.

We live in a "Get set" world and an increasing body of research suggests that our hyper-vigilant lifestyle is severely impacting the health of our bodies. Our daily stressors are constantly activating the HPA axis, priming our bodies for action. Unlike competitive athletes, the stresses in our bodies are not released from the pressures generated by our chronic fears and concerns. Almost every major illness that people acquire has been linked to chronic stress. (Segerstrom and Miller 2004; Kopp and Réthelyi 2004; McEwen and Lasky 2002; McEwen and Seeman 1999) Between 75 and 90 percent of primary-care physician visits have stress as a major contributing factor. (Atkinson 2000)

In a revealing study published in 2003 in *Science,* researchers considered why patients on SSRI antidepressants, such as Prozac or Zoloft, don't feel better right away. There is usually at least a two-week lag between starting the drugs and the time the patients feel they are getting better. The study found that depressed people exhibit a surprising lack of cell division in the region of the brain called the hippocampus, a part of the nervous system involved with memory. Hippocampal cells renewed cell division at the time patients first began to experience the mood-shifting effect of the SSRI drugs, weeks after the onset of the drug regimen. This study and others challenge the theory that depression is simply the result of a "chemical imbalance" affecting the brain's production of monoamine signaling chemicals, specifically serotonin. If it were as simple as that, the SSRI drugs would likely restore that chemical balance right away.

More researchers are pointing to the inhibition of neuronal growth by stress hormones as the source of depression. In fact, in chronically depressed patients, the hippocampus and the prefrontal cortex, the center of higher reasoning, are physically shrunken. A review of this study published in *Science* reported: "Overtaking the monoamine hypothesis in recent years has been the stress hypothesis, which posits that depression is caused when the brain's stress machinery goes into overdrive. The most prominent player in this theory is the hypothalamic-pituitary-adrenal (HPA) axis." (Holden 2003)

The HPA axis' effect on the cellular community mirrors the effect of stress on a human population. Picture a vibrant community back in the Cold War years, when the possibility of a nuclear attack by the Russians weighed heavily on Americans' minds. Like cells in a multicellular organism, the members of this Cold War society actively work at jobs that contribute to the community's growth and usually get along with each other. Factories are busy manufacturing, construction people are building new homes, grocery stores are selling food, and kids are in school learning their ABCs. The community is in a state of health and growth while its residents constructively interact toward a common goal.

Suddenly, the sound of an air raid siren rocks the town. Everyone stops working to run off, seeking the safety of bomb shelters. The harmony of the community is disrupted as individuals, acting only in support of their own survival, fight their way to a bomb shelter. After five minutes, the all-clear signal sounds. Residents return to their jobs and resume their lives in a growing community.

But what would happen if the sirens sound, the residents run into their air raid shelters, and there is no all-clear signal to release them? People would stay in their protective postures indefinitely. How long can they maintain their protection posture? The community eventually collapses in the face of dwindling food and water supplies. One by one even the strongest die because chronic stress is debilitating. A community can easily survive short-term stress, like an air raid drill, but when the stress goes on and on it results in cessation of growth and the breakdown of the community.

Another illustration of the influence of stress on a population is the story of the 9/11 tragedy. Up to the moment the terrorists attacked, the country was in a state of growth. Then immediately after 9/11, as the shock of the news spread to reach not just the people of New York but the entire nation, we experienced a threat to our survival. The impact of government proclamations stressing the continued presence of danger in the wake of the attack was like the influence of the adrenal signals. They shifted the members of the community from a state of growth to a state of protection. After a few days of this heart-stopping fear, the country's economic vitality was so compromised that the president had to intervene. To stimulate growth, the president repeatedly emphasized, "America is open for business." It took a while for the fears to subside and for the economy to rebound. However, the residual threats of terrorism are still debilitating the vitality of our country. As a nation we should look more carefully at how our fear of future acts of terrorism is undermining our quality of life. In some sense, the terrorists have already won since they have succeeded in frightening us into a chronic, soul-sapping protective mode.

I'd also like to suggest that you examine how your fears and the ensuing protection behaviors impact your life. What fears are stunting your growth? Where did these fears come from? Are they necessary? Are they real? Are they contributing to a full life? We'll deal more with these fears and where we got them in the next chapter on conscious parenting. If we can control our fears, we can regain control over our lives. President Franklin D. Roosevelt knew the destructive nature of fear. He chose his words carefully when he told the nation in the grips of the Great Depression and looming World War: "We have nothing to fear, but *fear* itself." Letting go of our fears is the first step toward creating a fuller, more satisfying life.

✳ ✳ ✳

Facebook was just a glint in a Harvard undergrad's eye when I first wrote this book. Ten years later I found a young woman (born decades after World War II) on Facebook who echoes Roosevelt's insight. Lupytha Hermin, an oft-cited artist who posts inspirational words with pictures on Facebook, offers this simple wisdom: "You know why it's hard to be happy—it's because we refuse to LET GO of the things that make us sad." I believe this may be one of the more important insights in learning how to control the negative effects of stress. People who live through negative and potentially devastating life experiences generally hold on to their fears and their stressful memories that compromise their health and longevity.

Though I believe it's hugely important to let go of our fear and stress-provoking memories, I also want to emphasize that not all stress is bad. In fact, there is a good form of stress known as eustress that has beneficial psychological and/or physical effects. Engaging in taxing exercise is one form of beneficial stress that enhances health and strengthens the body. (Of course, not all sports have positive effects—for Roman gladiators, the health benefits of exercise were no doubt completely offset by the negative psychological stress of knowing they would die if they lost!) Also, in some life-threatening situations, such as when we are physically attacked or when we are trying to manage a car spinning out of control, the stress response summoned up by the HPA axis can save one's life by calling forth almost super powers of strength and lightning-fast reactivity. That is the good side of the distress coin.

The bad side of the coin is when our bodies become chronically stimulated by misperceptions (unfortunately, our stress managing systems cannot distinguish whether a brain-directed response is derived from a real or an imagined fear) and sadness that we can't let go. This is dangerous to our health because we maintain the "get set" posture I mentioned earlier. In these situations, the stress response becomes chronic and results in a sustained release of the adrenal glucocorticoid hormone cortisol. Overstimulation of this glucocorticoid hormone is directly linked to significant damage and long-lasting functional changes in the brain.

For example, chronic stress-related illnesses such as PTSD are associated with alterations in the volume of the brain's gray and white matter; the region of the hippocampus associated with memory and emotions shrinks as does the amygdala, the brain's threat center. These observed brain alterations are believed to contribute to the creation of hardwired, stress-linked pathways between the hippocampus and amygdala that result in a vicious behavioral cycle of maintaining a constant fight-or-flight state of mind. If you've ever had a boss stuck in constant panic (the true boss from hell) or if you've ever had sleepless nights worrying rather than working on a deadline you fear you might not meet, you know how hardwired chronic stress can become. They are also damaging to children—children who develop chronic stress behaviors are more likely to experience learning impairment and psychological dysfunctions, such as anxiety and mood disorders, later in life. (Chetty, et al, 2014)

Chronic stress also depresses the immune system by impairing the function of glucocorticoid receptors normally used to inhibit or shut down inflammatory responses. This action conserves bodily energy to engage in a fight-or-flight response for what the mind perceives as life-threatening stress. Interfering with the behavior of these immune receptors results in a dysfunction referred to as Glucocorticoid Receptor Resistance (GCR) in which the duration and intensity of inflammatory responses increase, heightening the risk for asthma and other autoimmune diseases and encouraging the onset and progression of chronic inflammatory diseases such as cardiovascular disease, cancer, and type 2 diabetes. GCR is associated with people who experience chronic stress, such as parents of children with cancer, spouses of brain cancer patients, and persons reporting high levels of loneliness. (Cohen, et al, 2012)

Despite the risk of side effects, which range from dementia to premature death (Weich, et al, 2014), drugs are still biomedicine's go-to remedies for stress and anxiety. And that includes animals as well as people! In 2014, Iowa State University College of Veterinary Medicine researchers found that administering an anti-inflammatory drug to cattle before shipping them long distance relieved "transportation stress" as evidenced by stress biomarkers, including decreased cortisol. (Van Engen, et al, 2014)

But first-rate research about the effectiveness of nondrug remedies is accumulating, and that research fuels my relentless optimism that one day nondrug remedies will prevail. A number of studies in peer-reviewed journals have found that the "relaxation response," which researchers define as "the physiological and psychological state opposite to the stress or flight-or-flight response," alleviates symptoms of anxiety and many other disorders and also affects factors such as heart rate, blood pressure, oxygen consumption, and brain activity. And a groundbreaking study published in 2013 documented for the first time that this physiologic state of deep rest induced by practices like meditation, yoga, deep breathing, and prayer produces immediate changes in the expression of genes involved in immune function, energy metabolism, and insulin secretion, results that should impress even the most skeptical biomedical researchers. "Many studies have shown that mind/body interventions like the relaxation response can reduce stress and enhance wellness in healthy individuals and counteract the adverse clinical effects of stress in conditions like hypertension, anxiety, diabetes and aging," said pioneering mind-body researcher and co-author of the study Herbert Benson, M.D. "Now for the first time we've identified the key physiological hubs through which these benefits might be induced." (Bhasin, et al, 2013)

Other researchers are exploring the implications of Lupytha Hermin and Franklin D. Roosevelt's wisdom that I wrote about at the beginning of this chapter's update. Because holding on to our fears and our pain is a fundamental, underlying determinant for generating chronic stress behaviors, is it possible that love, the polar opposite of fear, could be an antidote for stress and its related diseases?

Using functional magnetic resonance imaging (fMRI) to monitor brain response, researchers at the University of Exeter addressed that question. They compared the brain responses of forty-two adults who viewed images of people expressing love and emotional support and images of "threatening" (angry or fearful) faces. The experimental group was briefly shown images of affection and emotional support that were then followed by threatening pictures; control subjects were only shown the threatening images.

In the control group, threatening images elicited high activity in the amygdala, the brain region that monitors threats. Participants in the experimental group, who were first presented with pictures expressing love and then shown the threatening images, showed no response in the amygdala to what should have been stress-provoking pictures. The researchers hypothesized that the neurological response to images of love suppresses the brain's threat mechanisms. Citing PTSD, which is characterized by hyper-vigilance to threatening information, lead researcher Anke Karl said: "These new research findings may help to explain why, for example, successful recovery from psychological trauma is highly associated with levels of perceived social support individuals receive. We are now building on these findings to refine existing treatments for PTSD to boost feelings of being safe and supported in order to improve coping with traumatic memories." (Norman, et al, 2014)

A number of studies confirm the healing influence that loving relationships and interactions have in our lives. As acclaimed neurobiologist Dr. Daniel Siegel reported in *The New York Times*, "Scientific studies of longevity, medical and mental health, happiness and even wisdom point to supportive relationships as the most robust predictor of these positive attributes in our lives across the life span." (Ackerman 2012)

Another study by University of Virginia neuroscientist James Coan assessed the role of social contact in regulating emotional responses in the face of various stressors. In Coan's assay, he gave an electric shock to the ankles and recorded the brain activity of sixteen women in three different situations: while holding the hand of a loved one, while holding the hand of an anonymous male, and while not holding a hand at all. The stress tests registered the women's anxiety before, and pain level during, the shocks. When not holding another person's hands, the women's anxiety and pain levels were elevated as expected. When holding their partner's hand, the negative influence of the shock was significantly reduced. Holding the hand of a stranger resulted in a much more limited reduction of the stress response. Strikingly, the degree of a woman's ability to reduce stress while holding hands with her loved one varied by the quality she attributed to her relationship: the higher a woman perceived the quality of her relationship, the lower were her scores on the anxiety and pain readings following the shock. (There was no such response variation associated with the holding of strangers' hands.) In a healthy relationship, holding your partner's hand is enough to lower blood pressure, ease stress responses, improve health, and diminish physical pain! (Coan, et al, 2006)

I am the last one to need convincing on that point because, as I described in my 2013 book, *The Honeymoon Effect,* my partner Margaret and I have created a wonderfully loving relationship that helps inoculate me from stress. Also, because of my history of bad relationships, I am a walking, talking example of the overwhelming science chronicling neuroplasticity (more in the next chapter) that has shown over and over that it is never too late—the brain is constantly undergoing structural and functional alterations in response to the influence of life and learning experiences.

When it comes to stress, the answer to the ever young and energetic Tina Turner's hit song, "What's Love Got to Do with It?" is: "Everything!" The creation of loving bonds assures the mind that when we are threatened, there will be somebody there to throw us a life preserver. That frees us, as it did me, from the need to observe our lives through filters of fear because we know we will be supported unconditionally. Individuals who are disconnected from social relationships and community, on the other hand, perceive they are alone and adrift in an ocean where no one ever comes to their aid.

Now I'd like to circle back to the beginning of this update, to Lupytha Hermin's message ("You know why it's hard to be happy—it's because we refuse to LET GO of the things that make us sad") and conclude with the story of Scarlett Lewis, a mother whose life embodies the wisdom of those words. Scarlett's youngest son, Jesse, was one of the twenty children murdered in the 2012 Sandy Hook Elementary

School massacre in Newtown, Connecticut. Brave Jesse helped save the lives of many of his classmates by encouraging them to run while he stayed behind to protect his teacher—both he and his beloved teacher were killed. (Lewis 2014)

Before going to school, in what may have been a premonition of the day's tragedy, six-year-old Jesse wrote on his home chalkboard, "Nurturing Healing Love." Working through her grief in the midst of the emotional devastation felt by all of the parents who lost children, Scarlett embraced Jesse's words and consciously chose a different way to manage her distress. While many parents vented their pain through anger, blame, and overwhelming grief, Scarlett went on an alternate path by deciding to consciously choose Love to come to terms with this heinous crime.

To send her message into the world, Scarlett founded the Jesse Lewis Choose Love Foundation (http://www.jesselewischooselove. org) whose stated mission is, "To create awareness in our children and our communities that we can choose love over anger, gratitude over entitlement, and forgiveness and compassion over bitterness." The foundation's goal is to help manifest a more peaceful and loving world. Scarlett's efforts in advancing Love to resolve the world's problems became her path to healing. Scarlett represents a living example of the powerful opportunity to heal offered in the final words of Jesus: "Father, forgive them, for they know not what they do."

CHAPTER 7

CONSCIOUS PARENTING:
Parents as Genetic Engineers

No doubt you've heard the seductive argument that once parents bestow their genes on their children, they take a back seat in their children's lives—parents need only refrain from abusing their children, feed and clothe them, and then wait to see where their preprogrammed genes lead them. This notion allows parents to continue with their "pre-kids lives"—they can simply drop their children off at daycare and with babysitters. It's an appealing idea for busy and/or lazy parents.

It's also appealing for parents like me, who have biological children with radically different personalities. I used to think that my daughters are different because they inherited different sets of genes from the moment of conception—a random selection process in which their mother and I had no part. After all, I thought, they grew up in the same environment (nurture), so the reason for their differences had to be genetic (nature).

The reality, I know now, is very different. Frontier science is confirming what mothers and enlightened fathers have known forever, that parents *do* matter, despite best-selling books that try to convince them otherwise. To quote Dr. Thomas Verny, a pioneer in the field of prenatal and perinatal psychiatry: "Findings in the peer-reviewed literature over the course of decades establish, *beyond any doubt,* that parents have overwhelming influence on the mental and physical attributes of the children they raise." (Verny and Kelly 1981)

And that influence starts, says Verny, not after children are born, but BEFORE children are born. When Verny first posited the notion that the influence of parents extends even to the womb in his landmark, 1981 book, *The Secret Life of the Unborn Child,* the scientific evidence was preliminary and the "experts" skeptical. Because scientists used to think that the human brain did not become functional until after birth, it was assumed that fetuses and infants had no memory and felt no pain. After all, noted Freud, who coined the termed "infantile amnesia," most people do not remember anything that happened to them before they were three or four years old.

However, experimental psychologists and neuroscientists are demolishing the myth that infants cannot remember—or for that matter learn—and along with it the notion that parents are simply spectators in the unfolding of their children's lives. The fetal and infant nervous system has vast sensory and learning capabilities and a kind of memory that neuroscientists call implicit memory. Another pioneer in pre- and perinatal psychology, David Chamberlain writes in his book *The Mind of Your Newborn Baby:* "The truth is, much of what we have traditionally believed about babies is false. They are not simple beings but complex and ageless—small creatures with unexpectedly large thoughts." (Chamberlain 1998)

These complex, small creatures have a pre-birth life in the womb that profoundly influences their long-term health and behavior. "The quality of life in the womb, our temporary home before we were born, programs our susceptibility to coronary artery disease, stroke, diabetes, obesity, and a multitude of other conditions in later life," writes Dr. Peter W. Nathanielsz in *Life in the Womb: The Origin of Health and Disease.* (Nathanielsz 1999) Recently, an even wider range of adult-related chronic disorders, including osteoporosis, mood disorders, and psychoses, have been intimately linked to pre- and perinatal developmental influences. (Gluckman and Hanson 2004; Shonkoff, et al, 2009)

Recognizing the role the prenatal environment plays in creating disease forces a reconsideration of genetic determinism. Nathanielsz writes: "There is mounting evidence that programming of lifetime health by the conditions in the womb is equally, if not more important, than our genes in determining how we perform mentally and physically during life. *Gene myopia* is the term that best describes the current all-pervasive view that our health and destiny throughout life are controlled by our genes alone. In contrast to the relative fatalism of gene myopia, understanding the mechanisms that underlie programming by the quality of life in the womb, we can improve the start in life for our children and their children."

The programming "mechanisms" Nathanielsz refers to are the epigenetic mechanisms, discussed earlier, by which environmental stimuli regulate gene activity. As Nathanielsz states, parents can improve the prenatal environment. In so doing they act as genetic engineers for their children. The idea that parents can transmit hereditary changes from their life to their children is, of course, a Lamarckian concept that conflicts with Darwinism. Nathanielsz is one of the scientists now brave enough to invoke the "L" word for Lamarck: "the transgenerational passage of characteristics by nongenetic means does occur. Lamarck was right, although transgenerational transmission of acquired characteristics occurs by mechanisms that were unknown in his day."

The responsiveness of individuals to the environmental conditions perceived by their mothers before birth allows them to optimize their genetic and physiologic development as they adapt to the environmental forecast. The same life-enhancing epigenetic plasticity of human development can go awry and lead to an array of chronic diseases in older age if an individual experiences adverse nutritional and environmental circumstances during fetal and neonatal periods of development. (Bateson, et al, 2004)

The same epigenetic influences also continue after the child is born because parents continue to influence their child's environment. In particular, fascinating new research is emphasizing the importance of good parenting in the development of the brain. "For the growing brain of a young child, the social world supplies the most important experiences influencing the expression of genes, which determines how neurons connect to one another in creating the neuronal pathways which give rise to mental activity," writes

Dr. Daniel J. Siegel in *The Developing Mind.* (Siegel 1999) In other words, infants need a nurturing environment to activate the genes that develop healthy brains. Parents, the latest science reveals, continue to act as genetic engineers even after the birth of their child.

Parental Programming: The Power of the Subconscious Mind

I'd like to tell you about how I—who put myself in the category of those who were *not* prepared to have children—came to question my ingrained assumptions about parenting. You won't be surprised to hear that I started my re-evaluation in the Caribbean, the place where my shift to the New Biology took root. My reassessment was actually inspired by an unlucky event, a motorcycle accident. I was on my way to present a lecture when I went off a curb at high speed. The bike wound up upside down. Luckily I was wearing a helmet because I sustained a major blow to my head when the bike hit the ground. I was unconscious for half an hour and for a while my students and colleagues thought I was dead. When I came to, I felt as if I had broken every bone in my body.

For the next few days I could hardly walk, and when doing so, I resembled a yelping version of Quasimodo. Every step was a painful reminder that "speed kills." As I creaked out of the classroom one afternoon, one of my students suggested that it might help if I visited his roommate, a fellow student, who was also a chiropractor. As I explained in the last chapter, I not only had never been to a chiropractor, I had been taught by my allopathic community to shun chiropractors as quacks. But when you're in that much pain and you're in an unfamiliar setting, you wind up trying things you would never consider in your cushier moments.

At the chiropractor's makeshift dormitory "office" I was introduced for the first time to kinesiology, popularly known as muscle testing. The chiropractor told me to hold out my arm and resist the downward pressure he applied to it. I had no problem resisting the light force he put on my arm. Then he asked me to hold out my arm and resist him again while I said, "My name is Bruce." Again, I had no trouble resisting him, but by now I was starting to think that the admonishments of my academic colleagues were right on the mark—"This is nuts!" Then, the chiropractor told me to hold out my arm and resist his pressure while saying earnestly, "My name is Mary." To my amazement, my arm flopped down, despite my strong resistance. "Now wait a minute," I said. "I must not have been resisting enough, try that again." So we did, and this time I concentrated even more forcefully on resisting. Nevertheless, after repeating, "My name is Mary," my arm sunk like a stone. This student, who was now *my* teacher, explained that when your conscious mind has a belief that is in conflict with a formerly learned "truth" stored in the subconscious mind, the intellectual conflict expresses itself as a weakening of the body's muscles.

To my astonishment, I realized that my conscious mind, which I exercised so confidently in academic settings, was not in control when I voiced an opinion that differed from a truth stored in the unconscious mind. My unconscious mind was undoing the best efforts of my conscious mind to hold up my arm when I claimed my name was Mary. I was amazed to discover that there was another "mind," another force that was co-piloting my life. More shocking was the fact that this hidden mind, the mind I knew little about (except theoretically in psychology) was actually more powerful than my conscious mind, just as Freud had claimed. All in all, my first visit to a chiropractor turned out to be a life-changing experience. I learned that chiropractors could tap into the body's innate healing power using kinesiology to target spinal misalignments. I was able to saunter out of that dorm feeling like a new man after a few simple, vertebral adjustments on the "quack's" table . . . all without the use of drugs. And most importantly, I was introduced to the "man behind the curtain," my subconscious mind!

As I left the campus, my conscious mind was awhirl over the implications of the superior power of my formerly hidden subconscious mind. I also coupled those musings with my study of quantum physics, which taught me that thoughts could propel behavior more efficiently than physical molecules. My subconscious "knew" that my name was not Mary and balked at my insistence that it was. What else did my subconscious mind "know," and how had it learned it?

To understand better what had happened in that chiropractor's office, I first turned to comparative neuroanatomy, which reveals that the lower an organism is on the Tree of Evolution, the less developed its nervous system and thus the more it relies on preprogrammed behavior (nature). Moths fly toward the light, sea turtles return to specific islands and lay their eggs on the beach at the appropriate time, and the swallows return to Capistrano on a specific date, yet, as far as we know, none of these organisms have any knowledge of why they engage in those behaviors. The behaviors are innate; they are genetically built into the organism and are classified as *instincts*.

Organisms higher in the Tree have more complexly integrated nervous systems headed by bigger and bigger brains that allow them to acquire intricate behavioral patterns through experiential learning (nurture). The complexity of this environmental learning mechanism presumably culminates with humans, who are at the top, or at least near the top, of the Tree. To quote anthropologists Emily A. Schultz and Robert H. Lavenda: "Human beings are more dependent on learning for survival than other species. We have no instincts that automatically protect us and find us food and shelter, for example." (Schultz and Lavenda 1987)

We do have, of course, behavioral instincts that are innate—consider the infant's instinct to suckle, to quickly move his hand away from fire, and to automatically swim when placed in water. Instincts are built-in behaviors that are fundamental to the survival of all humans, independent of what culture they belong to or what time in human history they were born. We are born with the ability to swim; infants can swim like graceful porpoises moments after they are born. But children quickly acquire a fear of water from their parents—observe the response of parents when their unattended child ventures near a pool or other open water. Children learn from their parents that water is dangerous. Parents must later struggle to teach Johnny how to swim. Their first big effort is focused on overcoming the fear of water they instilled in earlier years.

But through evolution, our *learned* perceptions have become more powerful, especially because they can override genetically programmed instincts. The body's physiological mechanisms (e.g., heart rate, blood pressure, blood flow/bleeding patterns, body temperature) are, by their nature, programmed instincts. However, yogis as well as everyday people using biofeedback can *learn* to consciously regulate these "innate" functions.

Scientists have focused on our big brains as the reason for our ability to learn such complex behavior. However, we should temper our enthusiasm for the big brain theory by considering that elephants, whales, and cetaceans (porpoises and dolphins) have a greater cerebral surface area packed into their craniums than we do.

And the findings of British neurologist Dr. John Lorber, highlighted in a 1980 article in *Science*—"Is Your Brain Really Necessary?"—also call into question the notion that the size of the brain is the most important consideration for human intelligence. (Lewin 1980) Lorber studied many cases of hydrocephalus ("water on the brain") and concluded that even when most of the brain's cerebral cortex, the brain's outer layer, is missing, patients can live normal lives. *Science* writer Roger Lewin quotes Lorber in his article:

> There's a young student at this university (Sheffield University) who has an IQ of 126, has gained a first-class honors degree in mathematics, and is socially completely normal. And yet the boy has virtually no brain . . . When we did a brain scan on him, we saw that instead of the normal 4.5 centimeter thickness of brain tissue between the ventricles and the cortical surface, there was just a thin layer of mantle measuring a millimeter or so. His cranium is filled mainly with cerebrospinal fluid.

Lorber's provocative findings suggest that we need to reconsider our long-held beliefs about how the brain works and the physical foundation of human intelligence. I submit in the Epilogue of this book that human intelligence can only be fully understood when we include spirit ("energy") or what quantum-physics-savvy psychologists call the "superconscious" mind. But for the moment, I'd like to stick to the conscious and subconscious minds, concepts that psychologists and psychiatrists have long grappled with. I'm grappling with it here to provide the biological foundation for conscious parenting as well as energy-based psychological healing methods.

Human Programming: When Good Mechanisms Go Bad

Let's go back to the evolutionary challenge for human beings, who have to learn so much so quickly to survive and become a part of their social community. Evolution has endowed our brains with the ability to rapidly download an unimaginable number of behaviors and beliefs into our memory. Ongoing research suggests that a key to understanding how this rapid downloading of information works is the brain's fluctuating electrical activity as measured by electroencephalograms. The literal definition of electroencephalograms (EEGs) is "electric head pictures." These increasingly sophisticated head pictures reveal a graded range of brain activity in human beings. Both adults and children display EEG variations that range from low frequency *delta* waves through high frequency *beta* waves. However, researchers have noted that EEG activity in children reveals, at every developmental stage, the predominance of a specific brain wave.

Dr. Rima Laibow in *Quantitative EEG and Neurofeedback* describes the progression of these developmental stages in brain activity. (Laibow 1999 and 2002) Between birth and two years of age, the human brain *predominantly* operates at the lowest EEG frequency, 0.5 to 4 cycles per second (Hz), known as *delta* waves. Though *delta* is their predominant wave activity, babies can exhibit periodic short bursts of higher EEG activity. A child begins to spend more time at a higher level of EEG activity characterized as *theta* (4-8 Hz) between two and six years of age. Hypnotherapists drop their patients' brain activity into *theta* because this low frequency brain wave puts them into a more suggestible, programmable state.

This gives us an important clue as to how children, whose brains are mostly operating at this frequency through six years of age, can download the incredible volume of information they need to thrive in their environment. The ability to process this vast quantity of information is an important neurologic adaptation to facilitate this information-intense process of enculturation. Human environments and social mores change so rapidly that it would not be an advantage to transmit cultural behaviors via genetically programmed instincts. Young children carefully observe their environment and download the worldly wisdom offered by parents directly into their subconscious memory. As a result, their parents' behavior and beliefs become their own.

Researchers at Kyoto University's Primate Research Institute have found that baby chimps also learn by simply observing their mothers. In a series of experiments over a period of two years, a mother was taught to identify the Japanese characters for a variety of colors. When the Japanese character for a specific color was flashed on a computer screen, the chimp learned to choose the right color swatch. Upon selecting the right color, the chimp received a coin that she could then put in a vending machine for a fruit treat. During her training process, she was holding her baby close. To the surprise of researchers, one day, as the mother was retrieving her fruit from the vending machine, the infant chimp activated the computer. When the character appeared on the screen, the baby chimp selected the correct color, received a coin, and then followed his mother to the vending machine. The astonished researchers were left to conclude that infants can pick up complex skills solely by observation and don't have to be actively coached by their parents. (Science 2001)

In humans as well, the fundamental behaviors, beliefs, and attitudes we observe in our parents become "hardwired" as synaptic pathways in our subconscious minds. Once programmed into the subconscious mind, they control our biology for the rest of our lives . . . or at least until we make the effort to reprogram them. Anyone who doubts the sophistication of this downloading should think about the first time your child blurted out a curse word picked up from you. I'm sure you noted its sophistication, correct pronunciation, its nuanced style, and context carrying your signature.

Given the precision of this behavior-recording system, imagine the consequences of hearing your parents say you are a "stupid child," you "do not deserve things," will "never amount to any-thing," "never should have been born," or are a "sickly, weak" person. When unthinking or uncaring parents pass on those messages to their young children, they are no doubt oblivious to the fact that such comments are downloaded into the subconscious memory as absolute "facts" just as surely as bits and bytes are downloaded to the hard drive of your desktop computer. During early development, the child's consciousness has not evolved enough to critically assess that those parental pronouncements were only verbal barbs and not necessarily true characterizations of "self." Once programmed into the subconscious mind, however, these verbal abuses become defined as "truths" that unconsciously shape the behavior and potential of the child through life.

At around the age of six, we become less susceptible to outside programming with the increasing appearance of higher frequency *alpha* waves (8-12 Hz). *Alpha* activity is equated with states of calm consciousness. While most of our senses, such as eyes, ears, and nose, observe the outer world, consciousness resembles a "sense organ" that behaves like a mirror reflecting back the inner workings of the body's own cellular community; it is an awareness of "self."

At around twelve years of age, the child's EEG spectrum begins to show sustained periods of an even higher frequency defined as *beta* waves (12-35 Hz). *Beta* brain states are characterized as "active or focused consciousness," the kind of brain activity used in reading this book. Recently, a fifth, higher state of EEG activity has been defined. Referred to as *gamma* waves (>35 Hz), this EEG frequency range kicks in during states of "peak performance," such as when pilots are in the process of landing a plane or a professional tennis player is engaged in a rapid-fire volley.

By the time children reach adolescence, their subconscious minds are chock-full of information that ranges from the knowledge of how to walk to the "knowledge" they will never amount to anything or the knowledge, fostered by loving parents, that they can do anything they set out to do. The sum of our genetically programmed instincts and the beliefs we learned from our parents collectively form the fundamental programs in the subconscious mind. Programs that can undo both our ability to keep our arm raised in a chiropractor's office and our best New Year's resolutions to stop sabotaging ourselves with drugs or food.

Again I go back to cells, which can teach us so much about ourselves. I've said many times that single cells are intelligent. But remember, when cells band together in creating multicellular communities, they follow the "collective voice" of the organism, even if that voice dictates self-destructive behavior. Our physiology and behavior patterns conform to the "truths" of the central voice, be they constructive or destructive beliefs.

I've described the power of the subconscious mind, but I want to emphasize that there is no need to consider the subconscious a scary, super-powerful, Freudian font of destructive "knowledge." In reality, the subconscious is an emotionless database of stored programs, whose function is strictly concerned with reading environmental signals and engaging in hardwired behavioral programs, no questions asked, no judgments made. The subconscious mind is similar to a programmable "hard drive" into which our life experiences are downloaded. The programs are functionally equivalent to hardwired stimulus-response behaviors. Behavior activating stimuli may be signals the nervous system detects from the external world and/or signals that arise from within the body such as emotions, pleasure, and pain. When a stimulus is perceived, it will automatically engage the behavioral response that was learned when the signal was first experienced. In fact, people who realize the automated nature of this playback response frequently admit to the fact that their "buttons have been pushed."

Before the evolution of the conscious mind, the functions of animal brains consisted only of those that we link with the subconscious mind. These more primitive minds were simple stimulus-response devices that automatically responded to environmental stimuli by engaging genetically programmed (instincts) or simple learned behaviors. These animals do not "consciously" evoke such behaviors, and in fact, may even be oblivious to them. Their behaviors are programmed reflexes, like the blink of an eye in response to a puff of air or the kick of a leg after tapping the knee joint.

The Conscious Mind: The Creator Within

The evolution of higher mammals, including chimps, elephants, cetaceans, and humans, brought forth a new level of awareness called "self-consciousness," or, simply, the conscious mind. The newer conscious mind is an important evolutionary advance. The earlier, subconscious mind is our "autopilot"; the conscious mind is our manual control. For example, if a ball comes near your eye, the slower conscious mind may not have time to be aware of the threatening projectile. Yet the subconscious mind, which processes some 20 million environmental stimuli per second versus forty environmental stimuli interpreted by the conscious mind in the same second, will cause the eye to blink. (Norretranders 1998). The subconscious mind, the most powerful information processor known, specifically observes both the surrounding world and the body's internal awareness, reads the environmental cues, and immediately engages previously acquired (learned) behaviors—all without the help, supervision, or even awareness of the conscious mind.

The two minds make a dynamic duo. Operating together, the conscious mind can use its resources to focus on some specific point, such as the party you are going to on Friday night. Simultaneously, your subconscious mind can be safely pushing the lawn mower around and successfully not cutting off your foot or running over the cat—even though you are not consciously paying attention to mowing the lawn.

The two minds also cooperate in acquiring very complex behaviors that can subsequently be unconsciously managed. Remember the first day you excitedly sat in the driver's seat of a car, preparing to learn how to drive? The number of things that had to be dealt with by the conscious mind was staggering. While keeping your eyes on the road, you had to also watch the rear and side view mirrors; pay attention to the speedometer and other gauges; use two feet for the three pedals of a standard shift vehicle; and try to be calm, cool, and collected as you drove past observing peers. It took what seemed to be a long time before all these behaviors were "programmed" into your mind.

Visualizing the information-processing powers of the conscious and subconscious minds. Consider that the image of Machu Picchu above is composed of 20 million pixel dots, each representing a BIT of information received by the nervous system. The powerful subconscious mind processes all this information in one second. How much of that incoming information enters the conscious mind? In the lower picture, the dot represents the total amount of information that is processed by the conscious mind in that same second. (Actually the dot is 10X more than enters consciousness. I had to enlarge it because it was barely visible.)

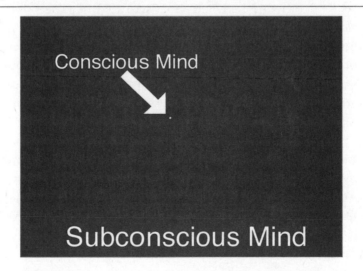

Today, you get in the car, turn the ignition on, and consciously review your shopping list as the subconscious mind dutifully engages all the complex skills you need to successfully navigate through the city—without even once having to think about the mechanics of driving. I know I am not the only one out there who has experienced this. You are driving and having a delightful discussion with the passenger sitting next to you. In fact, your consciousness gets so caught up in the conversation, that somewhere down the road it dawns on you that you haven't even paid attention to your driving for five minutes. After a momentary start, you realize that you are still on your side of the road and steadily moving along with the flow of traffic. A quick check of the rearview mirror reveals that you did not leave a wake of crumpled stop signs and smashed mailboxes. If you weren't consciously driving the car during that time, then who was? The subconscious mind! And how well did it do? Although you didn't observe its behavior, the subconscious mind apparently performed just as well as it was taught during your driver's education experience.

In addition to facilitating subconscious habitual programs, the conscious mind also has the power to be spontaneously creative in its responses to environmental stimuli. In its self-reflective capacity, the conscious mind can observe behaviors as they're being carried out. As a preprogrammed behavior is unfolding, the observing conscious mind can step in, stop the behavior, and create a new response. Thus the conscious mind offers us free will, meaning we are not just victims of our programming. To pull that off, however, you have to be fully conscious lest the programming take over, a difficult task, as anyone who's tried willpower can attest. Subconscious programming takes over the moment your conscious mind is not paying attention.

The conscious mind can also think forward and backward in time, while the subconscious mind is always operating in the present moment. When the conscious mind is busy daydreaming, creating future plans, or reviewing past life experiences, the subconscious mind is always on duty, efficiently managing the behaviors required at the moment, without the need of conscious supervision.

The two minds are truly a phenomenal mechanism, but here is how it can go awry. The conscious mind is the "self," the voice of our own thoughts. It can have great visions and plans for a future filled with love, health, happiness, and prosperity. While we focus our consciousness on happy thoughts, who is running the show? The subconscious. How is the subconscious going to manage our affairs? Precisely the way it was programmed. The behaviors managed by the subconscious mind when we are not paying attention may not be of our own creation because most of our fundamental behavioral programs were downloaded without question from observing other people. Because subconscious-generated behaviors are *not* generally observed by the conscious mind, many people are stunned to hear that they are just like their mom or their dad, the people who programmed their subconscious minds.

The learned behaviors and beliefs acquired from other people, such as parents, peers, and teachers, may not support the goals or desires of our conscious mind. The biggest impediments to realizing the successes of which we dream are the limitations programmed into the subconscious. These limitations not only influence our behavior, they can also play a major role in determining our physiology and health. As we've seen, the mind plays a powerful role in controlling the biological systems that keep us alive.

Nature did not intend the presence of the dual minds to be our Achilles heel. In fact, this duality offers a wonderful advantage for our lives. Consider it this way: what if we had conscious parents and teachers who served as wonderful life models, always engaging in humane and win-win relations with everyone in the community? If our subconscious mind were programmed with such healthy behaviors, we could be totally successful in our lives without ever being conscious!

The Subconscious Mind:
I Keep Calling and No One Answers

While the "thinking-self" nature of the conscious mind evokes images of a "ghost in the machine," there is no similar self-awareness operating in the subconscious mind. The latter mechanism is more akin to a jukebox loaded with behavioral programs, each ready to play as soon as appropriate environmental signals appear and press the selection buttons. If we don't like a particular song in the jukebox, how much yelling at or arguing with the machine will cause it to reprogram its playlist? In my college days, I saw many an inebriated student, to no avail, curse and kick jukeboxes that were not responsive to their requests. Similarly, we must realize that no amount of yelling or cajoling by the conscious mind can ever change the behavioral "tapes" programmed into the subconscious mind. Once we realize the ineffectiveness of this tactic, we can quit engaging in a pitched battle with the subconscious mind and take a more clinical approach to reprogramming it. Engaging the subconscious in battle is as pointless as kicking the jukebox in the hope that it will reprogram its playlist.

The futility of battling with the subconscious is a hard message to get across because one of the programs most of us downloaded when we were young is that "willpower is admirable." So we try over and over again to override the subconscious program. Usually such efforts are met with varying degrees of resistance because the cells are obligated to adhere to the subconscious program.

Tensions between conscious willpower and subconscious programs can result in serious neurological disorders. For me, a powerful image of why we should not challenge the subconscious comes from the movie *Shine*. In the movie, based on a true story, Australian concert pianist David Helfgott defies his father by going off to London to study music. Helfgott's father, a survivor of the Holocaust, programmed his son's subconscious mind with the belief that the world was unsafe, that if he "stood out" it might be life threatening. His father insisted he would be safe only if he stayed close to his family. In spite of his father's relentless programming, Helfgott knew that he was a world-class pianist who needed to break from his father to realize his dream.

In London, Helfgott played the notoriously difficult *Third Piano Concerto* of Rachmaninoff in a competition. The film shows the conflict between his conscious mind wanting success and his subconscious mind concerned that being visible, being internationally recognized, was life threatening. As he labors through the concerto, sweat pouring from his brow, Helfgott's conscious mind fights to stay in control, while his subconscious mind, fearful of winning, tries to take control of his body. Helfgott consciously forces himself to maintain control through the concerto until he plays the last note. He then passes out, overcome by the energy it took to battle his subconscious programming. For that "victory" over the subconscious, he pays a high price: when he comes to, he is insane.

Most of us engage in less dramatic battles with our subconscious mind as we try to undo the programming we received as children. Witness our ability to continually seek out jobs that we fail at, or remain in jobs we hate, because we don't "deserve" a better life.

Conventional methods for suppressing destructive behaviors include drugs and talk therapy. Newer approaches promise to change our programming, recognizing that there is no use "reasoning" with the subconscious tape player. These methods capitalize on the findings of quantum physics that connect energy and thought. In fact, these new modalities that reprogram previously learned behaviors can be collectively referred to as energy psychology, a burgeoning field based on the New Biology.

But how much easier would it be to be nurtured from the beginning of life so that you can reach your genetic and creative potential? How much better to become a conscious parent so that your children and their children will be conscious parents, making reprogramming unnecessary and making for a happier, more peaceful planet!

A Twinkle in Your Parents' Eyes: Conscious Conception & Conscious Pregnancy

You all know the expression, "When you were only a twinkle in your parents' eyes"—a phrase that conjures up the happiness of loving parents who truly want to conceive a child. It turns out it is also a phrase that sums up the latest genetic research suggesting that parents should cultivate that twinkle in the months before they conceive a child. That growth-promoting awareness and intention can produce a smarter, healthier, and happier baby.

Research reveals that parents act as genetic engineers for their children in the months before conception. In the final stages of egg and sperm maturation, a process called *genomic imprinting* adjusts the activity of specific groups of genes that will shape the character of the child yet to be conceived. (Surani 2001; Reik and Walter 2001) Research suggests that what is going on in the lives of the parents during the process of genomic imprinting has a profound influence on the mind and body of their child, a scary thought given how unprepared most people are to have a baby. Verny writes in *Pre-Parenting: Nurturing Your Child from Conception:* "It makes a difference whether we are conceived in love, haste, or hate and whether a mother wants to be pregnant . . . parents do better when they live in a calm and stable environment free of addictions and supported by family and friends." (Verny and Weintraub 2002) Interestingly, aboriginal cultures have recognized the influence of the conception environment for millennia. Prior to conceiving a child, couples ceremonially purify their minds and bodies.

Once the child is conceived, an impressive body of research is documenting how important parents' attitudes are in the development of the fetus. Again Verny writes: "In fact, the great weight of the scientific evidence that has emerged over the last decade demands that we reevaluate the mental and emotional abilities of unborn children. Awake or asleep, the studies show, they (unborn children) are constantly tuned in to their mother's every action, thought, and feeling. From the moment of conception, the experience in the womb shapes the brain and lays the groundwork for personality, emotional temperament, and the power of higher thought."

Now is the time to stress that the New Biology is *not* a return to the old days of blaming mothers for every ailment that medicine didn't understand—from schizophrenia to autism. Mothers and fathers are in the conception and pregnancy business together, even though it is the mother who carries the child in her womb. What the father does profoundly affects the mother, which in turn affects the developing child. For example, if the father leaves and the mother starts questioning her own ability to survive, his leaving profoundly changes the interaction between the mother and the unborn baby. Similarly, societal factors, such as lack of employment, housing, and healthcare or endless wars that pull fathers into the military, can affect the parents and thus the developing child.

The essence of conscious parenting is that both mothers and fathers have important responsibilities for fostering healthy, intelligent, productive, and joy-filled children. We surely cannot blame ourselves nor our parents for failures in our own or our children's lives. Science has kept our attention focused on the notion of genetic determinism, leaving us ignorant about the influence beliefs have on our lives and more importantly, how our behaviors and attitudes program the lives of our children.

Most obstetricians are also still uneducated about the importance of parental attitudes in the development of the baby. According to the notion of genetic determinism that they were steeped in as medical students, fetal development is mechanically controlled by genes with little additional contribution from the mother. Consequently, ob-gyns are only concerned with a few maternal prenatal issues: Is she eating well? Taking vitamins? Does she exercise regularly? Those questions focus on what they believe is the mother's principal role, the provision of nutrients to be used by the genetically programmed fetus.

But the developing child receives far more than nutrients from the mother's blood. Along with nutrients, the fetus absorbs excess glucose if the mother is diabetic and excess cortisol and other fight-or-flight hormones if the mother is chronically stressed. Research now offers insights into how the system works. If a mother is under stress, she activates her HPA axis, which provides fight-or-flight responses in a threatening environment.

Stress hormones prepare the body to engage in a protection response. Once these maternal signals enter the fetal blood stream, they affect the same target tissues and organs in the fetus as they did in the mother. In stressful environments, fetal blood preferentially flows to the muscles and hindbrain, providing nutritional requirements needed by the arms and legs and by the region of the brain responsible for life-saving reflex behavior. In supporting the function of the protection-related systems, blood flow is shunted from the viscera organs and stress hormones suppress forebrain function. The development of fetal tissue and organs is proportional to both the amount of blood they receive and the function they provide. When passing through the placenta, the hormones of a mother experiencing chronic stress will profoundly alter the distribution of blood flow in her fetus and change the character of her developing child's physiology. (Lesage, et al, 2004; Christensen 2000; Arnsten 1998; Leutwyler 1998; Sapolsky 1997; Sandman, et al, 1994)

At the University of Melbourne, E. Marilyn Wintour's research on pregnant sheep, who are physiologically quite similar to humans, has found that prenatal exposure to cortisol eventually leads to high blood pressure (Dodic, et al, 2002). Fetal cortisol levels play a very important regulatory role in the development of the kidney's filtering units, the nephrons. A nephron's cells are intimately involved with regulating the body's salt balance and consequently are important in controlling blood pressure. Excess cortisol absorbed from a stressed mother modifies fetal nephron formation. An additional effect of excess cortisol is that it simultaneously switches the mother's and the fetus's system from a growth state to a protection posture. As a result, the growth-inhibiting effect of excess cortisol in the womb causes the babies to be born smaller.

Suboptimal conditions in the womb that lead to low birthweight babies have been linked to a number of adult ailments that Nathanielsz outlines in his book *Life in the Womb*, (Nathanielsz 1999) including diabetes, heart disease, and obesity. For example, Dr. David Barker (ibid.) of England's University of Southampton has found that a male who weighs fewer than 5.5 pounds at birth has a 50 percent greater chance of dying of heart disease than one with a higher birth weight. Harvard researchers have found that women who weigh fewer than 5.5 pounds at birth have a 23 percent higher risk of cardiovascular disease than women born heavier. And David Leon (ibid.) of the London School of Hygiene and Tropical Medicine has found that diabetes is three times more common in 60-year-old men who were small and thin at birth.

The new focus on the influences of the prenatal environment extends to the study of IQ, which genetic determinists and racists once linked simply to genes. But in 1997, Bernie Devlin, a professor of psychiatry at the University of Pittsburgh School of Medicine, carefully analyzed 212 earlier studies that compared the IQs of twins, siblings, and parents and their children. He concluded that genes account for only 48 percent of the factors that determine IQ. And when the synergistic effects of mingling the mother and father's genes are factored in, the true inherited component of intelligence plummets even further, to 34 percent. (Devlin, et al, 1997; McGue 1997)

Devlin, on the other hand, found that conditions during prenatal development significantly impact IQ. He reveals that up to 51 percent of a child's potential intelligence is controlled by environmental factors. Previous studies had already found that drinking or smoking during pregnancy can cause decreased IQ in children, as can exposure to lead in the womb. The lesson for people who want to be parents is that you can radically shortchange the intelligence of your child simply by the way you approach pregnancy. These IQ changes are not accidents; they are directly linked to altered blood flow in a stressed brain.

In my lectures on conscious parenting, I cite research, but I also show a video from an Italian conscious parenting organization, Associazione Nazionale per l'Educazione Prenatale, which graphically illustrates the interdependent relationship between parents and their unborn child. In this video, a mother and father engage in a loud argument while the woman is undergoing a sonogram. You can vividly see the fetus jump when the argument starts. The startled fetus arches its body and jumps up, as if it were on a trampoline when the argument is punctuated with the shattering of glass. The power of modern technology, in the form of a sonogram, helps to lay to rest the myth that the unborn child is not a sophisticated enough organism to react to anything other than its nutritional environment.

Nature's Head Start Program

You may be wondering why evolution would provide such a system for fetal development that seems so fraught with peril and is so dependent on the environment of the parents. Actually, it's an ingenious system that helps ensure the survival of your offspring. Eventually, when the child is born, it is going to find itself in the same environment as its parents. Information, in the form of regulatory hormones and emotional chemicals derived from the mother's perception of her environment, transits the placenta and primes the prenate's physiology, preparing it to more effectively deal with future exigencies that will be encountered after birth. Nature is simply preparing that child to best survive in that environment. However, armed with the latest science, parents now have a choice. They can carefully reprogram their limiting beliefs about life before they bring a child into their world.

The importance of parental programming undermines the notion that our traits, both positive and negative, are fully determined by our genes. As we have seen, genes are shaped, guided, and tailored by environmental learning experiences. We have all been led to believe that artistic, athletic, and intellectual prowess are traits simply passed on by genes. But no matter how "good" one's genes may be, if an individual's nurture experiences are fraught with abuse, neglect, or misperceptions, the realization of the genes' potentials will be sabotaged. Liza Minnelli acquired her genes from her superstar mother Judy Garland and her father filmmaker Vincent Minnelli. Liza's career, the heights of her stardom, and the lows of her personal life are scripts that were played out by her parents and downloaded into her subconscious mind. If Liza had the same genes but was raised by a nurturing Pennsylvania Dutch farming family, that environment would have epigenetically triggered a different selection of genes. The genes that enabled her to pursue a successful entertainment career would have likely been masked or inhibited by the cultural demands of her agrarian community.

A wonderful example of the effectiveness of conscious parenting programming is superstar golfer Tiger Woods. Although his father was not an accomplished golfer, he made every effort to immerse Tiger in an environment that was rich with opportunities to develop and enhance the mindset, skills, attitudes, and focus of a master golfer. No doubt, Tiger's success is also intimately connected with the Buddhist philosophy that his mother contributed. Indeed, genes are important—but their importance is only realized through the influence of conscious parenting and the richness of opportunities provided by the environment.

Conscious Mothering and Fathering

I used to close my public lectures with the admonition that we are personally responsible for everything in our lives. Such a closure did not make me popular with the audience. That responsibility was too much for many people to accept. After one lecture, an older woman in the audience was so distressed by my conclusion that she brought her husband backstage and in tears vehemently contested my conclusion. She did not want any part of some of the tragedies she had experienced. This woman convinced me that my summary conclusion had to be modified. I realized that I didn't want to contribute to foisting blame, shame, and guilt on any individual. As a society, we are too apt to wallow in guilt or scapegoat others for our problems. As we gain insights over a lifetime, we become better equipped to take charge of our lives.

After some deliberation, this woman from the audience happily accepted the following resolution: you are personally responsible for everything in your life, *once you become aware* that you are personally responsible for everything in your life. One cannot be "guilty" or be "blamed" for being a poor parent unless one was already aware of the above-described information and disregarded it. Once you become aware of this information, you can begin to apply it to reprogram your behavior.

And while we're on the subject of myths about parenting, it is absolutely not true that you are the same parent for all of your children. Your second child is not a clone of the first child. The same things are not happening in your world that happened when the first child was born. As I said above, I once thought that I was the same parent for my first child as I was for my very different second child. But when I analyzed my parenting, I found that was not true. When my first child was born, I was at the beginning of my graduate school training, which was for me a difficult transition fraught with a high workload accompanied by high insecurity. By the time my second daughter was born, I was a more confident, more accomplished research scientist ready to start my academic career. I had more time and more psychic energy to parent my second child and to better parent my first daughter, who was now a toddler.

Another myth I'd like to address is that infants need lots of stimulation in the form of black-and-white flash cards or other learning tools marketed to parents to increase the intelligence of their children. Michael Mendizza and Joseph Chilton Pearce's inspiring book *Magical Parent-Magical Child* makes it clear that *play* not programming is the key to optimizing the learning and performance of infants and children. (Mendizza and Pearce 2001) Children need parents who can playfully foster the curiosity, creativity, and wonder that accompanies their children into the world.

Obviously, what humans need is nurture in the form of love and the ability to observe older humans going about their everyday lives. When babies in orphanages, for example, are kept in cribs and only provided with food but not one-on-one smiles and hugs, they develop long-lasting developmental problems. One study of Romanian orphans by Mary Carlson, a neurobiologist at Harvard Medical School, concluded that the lack of touching and attention in Romanian orphanages and poor-quality day care centers stunted the children's growth and adversely affected their behavior. Carlson, who studied sixty Romanian children from a few months to three years of age, measured their cortisol levels by analyzing samples of saliva. The more stressed a child was, as determined by the higher than normal levels of cortisol in its blood, the poorer the outcome for the child. (Holden 1996)

Carlson and others have also done research on monkeys and rats demonstrating crucial links among touch, the secretion of the stress hormone cortisol, and social development. Studies by James

W. Prescott, former director of the National Institutes of Health's section on Human Health and Child Development, revealed that newborn monkeys deprived of physical contact with their mothers or social contact with others develop abnormal stress profiles and become violent sociopaths. (Prescott 1996 and 1990)

He followed up these studies with an assessment of human cultures based on how they raise their children. He found that if a society physically held and loved its children and did not repress sexuality, that culture was always peaceful. Peaceful cultures feature parents who maintain extensive physical contact with their children, such as carrying their babies on their chests and backs throughout the day. In contrast, societies that deprive their infants, children, and adolescents of extensive touch are inevitably violent in nature. One of the differences between populations is that many of the children not receiving touch suffer from somatosensory affective disorder. This disorder is characterized by an inability to physiologically suppress surging levels of stress hormones, a precursor to violent episodes.

These findings provide insight into the violence that pervades the United States. Rather than endorsing physical closeness, our current medical and psychological practices often discourage it. From the unnatural intervention of medical doctors in the natural process of birthing, for example, separating the neonate for extensive periods from the parents into distant nurseries, and the advising of parents not to respond to their babies cries for fear of spoiling them. Such practices, presumably based upon "science," undoubtedly contribute to the violence in our civilization. The research regarding touch and its relationship to violence is described in full at the following website: www.violence.de.

But what about the Romanian children who come out of deprived backgrounds and become what one researcher called "the resilient wonders"? Why do some children thrive despite their backgrounds? Because they have "better" genes? By now you know that I don't believe that. More likely, the birth parents of these resilient wonders provided a more nurturing pre- and perinatal environment as well as good nutrition at crucial points in the child's development.

The lesson for adoptive parents is that they should not pretend their children's lives began when they came into their new surroundings. Their children may have already been programmed by their birth parents with a belief that they are unwanted or unlovable. If more fortunate, they may have, at some crucial age in their development, received positive, life-affirming messages from their caretakers. If adoptive parents are not aware of pre- and perinatal programming, they may not deal realistically with post-adoption issues. They may not realize that their children did not come to them as a "blank slate" any more than newborns come into the world as blank slates, unaffected by their nine months in their mother's womb. Better to recognize that programming and to work, if necessary, to change it.

For adoptive and nonadoptive parents alike, the message is clear: Your children's genes reflect only their potential, not their destiny. It is up to you to provide the environment that allows them to develop to their highest potential.

Notice I do not say that it is up to parents to read lots of parenting books. I've met lots of people who are intellectually attracted to the ideas I present in this book. But intellectual interest is not enough. I tried that myself. I was intellectually aware of everything in this book, but before I made the effort to change, it made no impact on my life. If you simply read this book and think that your life and your children's lives will change, you're doing the equivalent of accepting the latest pharmaceutical pill, thinking it will "fix" everything. No one is fixed until they make the effort to change.

Here is my challenge to you. Let go of unfounded fears and take care not to implant unnecessary fears and limiting beliefs in your children's subconscious minds. Most of all, do not accept the fatalistic message of genetic determinism. You can help your children reach their potential and you can change your personal life. You are not "stuck" with your genes.

Take heed of the growth and protection lessons from cells and shift your lives into growth whenever possible. And remember that for human beings the most potent growth promoter is not the fanciest school, the biggest toy, or the highest-paying job. Long before cell biology and studies of children in orphanages, conscious parents and seers like Rumi knew that for human babies and adults the best growth promoter is love.

> A lifetime without Love is of no account
> Love is the Water of Life
> Drink it down with heart and soul

✳ ✳ ✳

Since I wrote the first edition of *The Biology of Belief*, research has continued to accumulate that parents do indeed serve as genetic engineers. One study I like to cite because though I hate brussels sprouts, I don't think I was genetically fated to hate them, is about mothers who ate greens during pregnancy or when breast-feeding. Instead of spitting out brussels sprouts as I would have, their babies downed them. "If mothers want their babies to learn to like to eat vegetables, especially green vegetables, they need to provide them with opportunities to taste these foods," Julie Mennella of the Monell Chemical Senses Center, who conducted the study, told *The Sunday Times*. (Leake 2007)

If your heart is sinking because you didn't eat any vegetables during pregnancy or if it already sank when you read earlier in the chapter that stressed-out moms pass their high levels of cortisol on to their babies and that you may have inadvertently negatively programmed your children's subconscious minds, I have to reiterate what I said earlier: DO NOT feel blame, shame, or guilt!!! These terms only legitimately apply when you have a proper understanding that something is harmful, and then *knowingly* choose to engage in it anyway. You cannot be "guilty" of bad parental behavior if you had no awareness or understanding of the implications of that behavior on your child's development.

Though research corroborating the importance of conscious parenting should never serve as a rationale for wallowing in guilt, I confess that it took me a while to put that into practice. When I first started to understand the crucial role parents play in their children's healthy development, I struggled with guilt, so much so that I felt the need to apologize to my daughters for my less than stellar parenting skills. I sat both of my young daughters down and told them that I feared my failures would screw up their lives. My oldest daughter responded with the equivalent of the proverbial teenager's "whatever" (to her credit as an adult, she later acknowledged that she was truly listening in spite of her feigned indifference); my younger daughter responded by carefully considering the evidence. While at the time they reacted differently, they both developed into wonderful adults and, in the last decade, loving mothers who practice conscious parenting, which goes to show that parents, such as myself, do not have to be perfect!

We humans, it turns out, are wonderfully resilient creatures, which you should keep in mind not only if you consider yourself a less than perfect parent but also if you were, like most of us, the recipient of dysfunctional programming when you were young. Though neuroscientists once thought our brains were fixed at adolescence, it is now an established fact that the wiring of the brain is "plastic," which means it can be rewired even into adulthood.

One of my favorite examples of neuroplasticity research is a University College London study of London's cab drivers. For 150 years, aspiring cab drivers have had to pass a test to prove they can successfully navigate London's notoriously labyrinthine streets and alleyways without a map, a task that takes on average more than two years. During that time, researchers found, the section of the drivers' hippocampi that is associated with spatial knowledge becomes bigger than average (the hippocampi of forty-year veteran London taxi drivers are even bigger!) "The change in hippocampus structure changed to accommodate their huge amount of navigating experience." (Maguire, et al, 2000)

More relevant to this conscious parenting chapter is an ongoing study of Romanian orphans that found that some of the debilitating effects of early deprivation can be addressed with appropriate nurturing, especially if it is provided within a critical period of development. As I described earlier, under the harsh leadership of Nicolae Ceausescu in the 1960s, horrific conditions prevailed in the country's orphanages, where a lack of socialization and communication between caregivers and institutionalized children resulted in profound behavioral dysfunctions, including a high incidence of autism and feral behavior. In this study, measurement of the EEG brain activity of institutionalized children showed much weaker signals than similarly aged children from the general population. However, when institutionalized infants were adopted into loving families before the age of two, their EEG patterns and behavior were no different than those of typical children by the time they reached eight years old. (Bhattacharjee 2015)

It also bears repeating that it's not too late as an adult to overcome your own negative programming by accessing your subconscious mind using many different processes including hypnosis, habituation (repetitive use of new behaviors), cognitive behavioral therapy, and a variety of rapid-change energy psychology modalities (listed on my website, www.brucelipton.com, under Resources). Given the changes I've brought about in my life using energy psychology and my contention that the medical industry is stuck in a pre-quantum-physics time warp, it should be no surprise that I believe that energy psychology options and lifestyle changes are preferable and overall more effective than pharmaceutical drugs.

Nevertheless, the drugging of America continues—one Mayo Clinic study concluded that nearly seven out of ten Americans are taking prescription drugs, and a whopping 20 percent take five or more! (Zhong, et al, 2013) This includes millions of courses of antibiotics that have spawned life-threatening, drug-resistant microbes. To put the situation into context, in Sweden, another developed country, the outpatient antibiotics prescription rate is less than half that of the U.S.: 388 per thousand versus 833. (Blaser 2014) The Centers for Disease Control and Prevention, warning that antibiotics should be used only for "critical situations," has concluded that "50% of all the antibiotics prescribed for people are not needed or are not optimally effective as prescribed." (CDC 2013) Most disturbing and most relevant for this chapter is the number of antibiotics being prescribed for children. In 2010 alone American children received 41 million courses of antibiotics, many of them to treat viral infections for which antibiotics don't work. "Most kids don't need them," writes Dr. Martin J. Blaser in his book *Missing Microbes: How the Overuse of Antibiotics Is Fueling Our Modern Plagues,* which I talked about in Chapter 1. (Blaser 2014) Researchers like Blaser are only beginning to understand the unintended side effects for children, including the decreasing of the diversity of our children's microbiome that makes them and likely future generations more susceptible to chronic disease.

Despite those warnings about antibiotics and despite the underperformance of statins that I talked about in Chapter 3, in 2008 the American Academy of Pediatrics (AAP) suggested that pediatricians consider prescribing *more* drugs in the form of statins for children before studies on their impact on human development are conducted! Acknowledging that childhood obesity has reached "epidemic" proportions, an epidemic that brings with it a higher risk of developing type 2 diabetes, hypertension, and cardiovascular disease, the AAP advised physicians to inform parents that they should nourish children using strict dietary guidelines, encourage them to engage in more physical activities, *and* "consider" prescribing cholesterol-lowering statin drugs for at-risk children as young as eight years old! (Daniels, et al, 2008) While this suggestion of a lifetime prescription for statins is morally offensive to me, consider it from the perspective of the massive, corporate financial windfall that pharmaceutical companies realize when they open up fertile new drug markets. In the United States, corporations are recognized as "people" by law; unfortunately, they often lack the most important of human characteristics: compassion and morality.

I am happy to report, however, that a year after the AAP suggested physicians consider cholesterol-lowering drugs for young children, the *Journal of the American Medical Association* (Shonkoff, et al, 2009), citing mounting epigenetic research that traces health problems back to fetal and early childhood stress, suggested a different approach. The *JAMA* article concluded that "new interventions to reduce significant stress in early childhood may be a more appropriate strategy for preventing adult heart diseases than the off-label administration of statins to school-aged children." Yeah!

The *JAMA* research cited points to a new way of thinking about obesity and subsequent cardiac dysfunctions, not as physiological disturbances, but as a consequence of early life adversities (aka the environment!), including, of course, neglectful or abusive parenting. Not long after conception and continuing into early infancy, immature organisms (i.e., children) read and respond to key environmental experiences, learning throughout the process to adapt by engaging in protective behaviors that ensure survival in a world characterized by high levels of stress and instability. Abusive developmental experiences specifically program a child's subconscious mind with vital protection behaviors to cope for a life in a dangerous world. As I explained in this chapter, these programmed responses to environmental adversity are essential and generally protective because they ensure short-term survival. However, persistent activation, mediated by chronic or overwhelming abuse, can cause a variety of pathogenic disturbances.

Statistical correlations reveal a direct connection between traumatic childhood experiences and a wide variety of health issues, including obesity, coronary artery disease, chronic pulmonary disease, cancer, alcoholism, depression, drug abuse, mental health problems, and teen pregnancies. Interference with developing behavioral processes can result in an adverse impact on adult health in either of two ways: (1) by inflicting cumulative damage over time (Gunnar and Quevedo 2007) or (2) by embedding destructive behaviors in the young mind that are only activated in adult life situations. (McEwen and Gianaros 2010) In either case, there can be a lag time of years before early environmental disturbances manifest as disease. For example, depressed adults with a history of childhood abuse are twice as likely to develop cardiovascular disease than depressed individuals with no history of juvenile maltreatment. (Shonkoff, et al, 2009)

In addition to the influence of psychosocial experiences in causing disease, environmental chemical toxins offer an alternative pathway to the same result. Developing organisms are extremely sensitive to the hormone-like activity of these chemical mimics. Today's technology is heavily contaminating the environment with many behavior-disturbing, endocrine-disrupting chemicals (EDCs) that include dioxin, phthalates, agricultural pesticides, polychlorinated biphenyls (PCBs), industrial solvents, pharmaceuticals, and heavy metals. Exposure to these EDCs that have estrogenic, antiestrogenic, and antiandrogenic properties has been shown to perturb the same stress pathways that provoke a disease response. (Vaiserman 2014)

These chemicals are entering the environment at the same time, researchers fear, that children's immune systems have been weakened not just by overuse of antibiotics but also by modern birth practices. For example, the number of C-section births has skyrocketed even though epidemiological studies have associated Cesarean section delivery with a greater risk for asthma, type 1 diabetes, obesity, and celiac disease in later life. British researchers found, for example, that the odds of being overweight or obese are 26 percent higher for babies born by Caesarean section. (Wong 2014) A recent Swedish study published in the *American Journal of Obstetrics & Gynecology* found epigenetic alterations in the stem cells of babies born via C-sections, though whether these changes are long lasting was beyond the scope of the study. The results found specific epigenetic differences between the groups in almost 350 DNA regions, including genes known to be involved in controlling metabolism and immune defense. (Almgren, et al, 2014) Another study by Swedish and Scottish researchers found that babies born via C-section, who do not have the benefit of acquiring all-important microbes when they travel through their mother's vagina, have a smaller and less diverse population of the phylum Bacteroidetes during the first two years, bacteria that aid in protection against allergies. (Jakobsson 2014)

The shift away from breast-feeding is also proving to be less than optimal for babies. The downgrading of breast milk started at the onset of the Industrial Revolution when job requirements demanded that urban mothers forgo breast-feeding. Between 1900 and 1960, negative social attitudes toward breast-feeding and the development of infant formula resulted in a significant decline in the number of children raised on breast milk. But despite attempts to convince the public that milk is milk, be it human breast milk, cow's milk, or formula, nothing could be further from the truth. Artificial formulas do not contain the powerful energy resources or immune protection found in mother's milk. In fact, a nutritional imbalance in synthetic formula feeding is associated with deaths from diarrhea in infants in both developing and developed countries. (Victora, et al, 1989)

Nature has evolved a chemical composition for breast milk that is specifically formulated to support the growth and health of human babies that has yet to be replicated. Breast milk contains antibodies that immunologically protect the baby until its own immune system becomes functional; to compensate for a baby's immature and non-functioning immune system, human milk contains vitally important immunoglobulin A (IgA) antibodies. These antibodies, which defend against infections in the infant's respiratory, digestive, and urogenital systems, provide passive immunity until the infant's own immune system can actively produce antibodies. Mother's milk is also a significant source of the bacteria that are proving to be crucial not only for a healthy digestive tract but also for a healthy immune system. Breast milk provides the highest energy and most concentrated source of lipids that are required to build the brain's structure—there is simply no substitute that can match complex maternal lipids as a readily assimilable, "high-octane" energy source.

And there is no doubt that an infant's brain needs lots of energy! Because humans have the largest brain size relative to body size among mammals, the energy demands of the human brain match those of the muscles of a marathon runner. A baby's brain doubles in size during the first year; in year two it reaches 80 percent of the size of an adult brain. In this rapid growth process, brain tissues burn up 60 percent of the baby's energy resources. (Gibbons 1998)

Most animals acquire their energy through their digestive tracts. It is a huge energy investment for herbivores like cows to extract nutrition (energy) from grass. That's the reason those mammals have the largest digestive systems relative to body size. The human digestive tract on the other hand is unusually small, only 60 percent of the size expected for a similar-sized primate. Paleoanthropologist Leslie C. Aiello believes that the large size of the human brain evolved through an ancestral dietary shift from a heavily vegetarian to a more energy dense and easily digestible diet that included meat. (Gibbons 1998) He also subscribes to the "maternal investment" hypothesis to explain where the energy comes from to fuel a baby's large brain. That hypothesis suggests that mothers provide for the brain's massive energy usage by transporting energy-dense nutrients through the placenta prenatally and later through breast milk until the child is about three years of age and its brain is approximately 85 percent of its adult size. (Gibbons 1998)

Breast-feeding also promotes mother-child attachment and decreases stress. The pleasure and bonding the infant derives from breast-feeding enhances the population of stress receptors found in the brain and can diminish cortisol activity by simultaneously releasing hormones in both the mother and child that create a loving bond between them. During breast-feeding, the brains of both the mother and the infant simultaneously release the pleasure hormone dopamine and the bonding hormone oxytocin. These secretions create a strong, loving bond between mother and child, especially when accompanied by eye-to-eye gazing. While an infant is suckling and receiving milk, it is also being inoculated with friendly bacteria found on the surface of the mother's breast.

You might think it's easy for me as a man to tout the maternal investment hypothesis and natural childbirth because I've never experienced the pain of childbirth! I would be the first to agree and I've never gone to the empathy-inducing lengths that some men have in which they have electrodes taped to their stomachs to simulate the pain of natural childbirth. (Burkitt 2014) I'm not presenting this research to induce guilt and to dictate women's decisions, but instead to stress that we should look at the modern medical industry's "improvements" with extreme skepticism. Whether it's messing with evolution in the form of decreasing the diversity of our microbiome, creating formula that doesn't hold a candle to mother's milk, promoting unnecessary C-sections, overprescribing antibiotics, or substituting parent-child interaction with a babysitting television screen, I think it's clear that we've gone astray by tampering with evolutionary mechanisms we don't fully understand.

Nevertheless I'd like to end this chapter on a positive note about how research into how humans acquire language is leading to better informed, conscious parents. Though there has been a cultural misunderstanding that a baby's brain is not developed enough to learn and comprehend language, nothing could be further from the truth. The acquisition of language plays a fundamental role in exercising an infant's brain and shaping its organization, neural connectivity, and intelligence. Research on the fetal brain's ability to acquire and download environmental experiences in the womb reveal that the nervous system's sensory input mechanisms, such as hearing, develop long before the system's motor outputs—in this case, coordinated muscular control needed for speech. Consequently, the brain's potential to learn and understand language is not dependent on the infant's ability to speak.

University of Washington neuroscientist Patricia Kuhl's studies make it clear that infants need adults in their lives to learn language effectively. Though weary parents may sometimes wish they could buy a video that has the same effect as engaging one-on-one with their child, it turns out there is no substitute for parents *interacting* with their infants and young children. In one study, Kuhl exposed nine-month-old children from English-speaking families to Mandarin. Some of the children interacted with native Chinese-speaking tutors, who played with them and read to them. A second group saw and heard the same Mandarin-speaking tutors through a video presentation. The third group heard only the audio track. After twelve sessions, the children were tested on their ability to discriminate between similar phonetic sounds in Mandarin. Children exposed to the language through *human interactions* were able to discriminate between similar Mandarin sounds as well as native listeners. Children in the other two groups learned nothing. The result of this and other studies has led Kuhl to propose the social gating hypothesis: the idea that an *interactive* social experience, i.e., conversation, is a portal to linguistic, cognitive, and emotional development. (Kuhl 2011)

Child psychologists Betty Hart and Todd Risley learned the same thing when they recorded hundreds of hours of interactions between children and adults in forty-two families from across a wide socioeconomic spectrum and assessed the children's development from nine months to three years. Children in well-to-do families, whose parents were typically college-educated professionals, heard an average of 2,153 words an hour spoken to them. In contrast, the children of low-income families heard an average only 616 words per hour. By their third birthday, the children in well-to-do families heard *30 million more words* than economically deprived children and the amount of conversation parents had with their infants was directly proportional to IQ test scores assessed at three years of age and the performance in school of these children at ages nine and ten. (Hart and Risley 2003)

The exciting part is that Hart and Risley's research has spawned conscious parenting initiatives thanks to technology in the form of LENA (Language Environment Analysis) devices. LENA devices work like pedometers except they keep track of words rather than steps. The Thirty Million Words Initiative in Chicago is making LENA devices available to parents so they can track the numbers of words they expose their children to. After six weeks, researchers in Chicago found a 32 percent increase in the number of words the children heard. Says Dr. Dana Suskind, Director of the Thirty Million Words Initiative: "Every parent has the ability to grow their children's brain and impact their future." (Suskind 2013)

Evolution has created an important role for parents and guardians and—now that I'm a grandfather several times over, I have to add grandparents—loving adults who play a fundamental role in shaping the future of the children they raise and their children's children's future as well. Each of us must stop to consider how negative programming impacts children's futures, for we are helping to shape the fate of human civilization. Progress is all too slow for those of us who have focused on parents' role as genetic engineers for so long, but whether it's the Thirty Million Words Initiative that teaches parents how to improve their children's language skills or *JAMA* arguing for a nonstatin approach to improving children's health or medical professionals advocating a reduction in the use of microbiota-depleting antibiotics, it's an advancement toward the day when every human adult has the big picture in mind—no matter who raises a child, that child's behavior will influence the evolution of us all.

EPILOGUE

SPIRIT AND SCIENCE

The most beautiful and profound emotion we can experience is the sensation of the mystical. It is the power of all true science.
— Albert Einstein

We've come a long way since Chapter 1, when I faced my panicked medical students and started my journey to the New Biology. But throughout the book I have not strayed far from the theme I introduced in the first chapter—that smart cells can teach us how to live. Now that we're at the end of the book, I'd like to explain how my study of cells turned me into a spiritual person. I also want to explain why I am optimistic about the fate of our planet, though I concede that optimism is sometimes hard to maintain if you read the daily newspaper.

I've specifically separated my discussion of Spirit and Science from the preceding chapters of the book by entitling this section the Epilogue. An epilogue is generally a short section at the end of the work that details the fate of its character . . . in this case *moi.* When the awareness that prompted this book first came into my head twenty years ago, I saw something in it that was so profound it immediately transformed my life. In the first instant of my big "aha," my brain was reveling in the beauty of the newly envisioned mechanics of the cell membrane. A few heartbeats later I was overtaken by a joy that was so deep and wide, my heart ached and tears flowed from my eyes. The mechanics of the New Science provided insights that implied the existence of our spiritual essence and our immortality. For me, however, the conclusions were so unambiguous that I instantly went from nonbeliever to believer.

I know that for some of you the conclusions I am going to present in this section are too speculative. Conclusions drawn in the previous chapters of the book are based upon a quarter of a century of studying cloned cells and are grounded in the astonishing new discoveries that are rewriting our understanding of the mysteries of life. The conclusions I offer in this epilogue are also based upon my scientific training—they do not spring from a leap of religious faith. I know conventional scientists may shy away from them because they involve the influence of invisible, matter-shaping energy fields that many refer to as Spirit. I am confident in presenting them for two reasons.

One reason is a philosophical and scientific rule called Occam's razor. Occam's razor holds that when several hypotheses are offered to explain a phenomenon, the simplest hypothesis that accounts for most of the observations is the most likely hypothesis and should be considered first. The new science of the magical mem-Brain in conjunction with the principles of quantum physics offer the simplest explanation that accounts not only for the science of allopathic medicine but also for the philosophy and practice of complementary medicine and spiritual healing. Also, after so many years of personally applying the science I have outlined in this book, I can attest to its power to change lives.

However, I concede that while science led me to my euphoric moment of insight, the experience resembled instantaneous conversions described by mystics. Remember the biblical story of Saul who was knocked off his horse with a lightning bolt? For me, there was no lightning bolt that came forth from the Caribbean skies. But I ran wild-eyed into the medical library because of a flash of life-changing insight concerning the nature of the cell membrane that was "downloaded" into my awareness in the wee hours of the morning. In assessing the beauty and elegance of the membrane's mechanics, I was drawn to the conclusion that we are immortal, spiritual beings who exist separately from our bodies. I had heard an undeniable inner voice informing me that I was leading a life based not only on the false premise that genes control biology but also on the false premise that we end when our bodies die. I had spent years studying molecular control mechanisms within the physical body and at that astounding moment came to realize that the protein "switches" that control life are primarily turned on and off by signals from the environment . . . the Universe.

You may be surprised that it was science that led me to that moment of spiritual insight. In scientific circles, the word "spirit" is as warmly embraced as the word "evolution" is in fundamentalist circles.

As you know, spiritualists and scientists approach life in vastly different ways. When life is out of whack for spiritualists, they beseech God or some other invisible force for relief. When life is out of whack for scientists, they run to the medicine cabinet for a chemical. It is only with a drug like Rolaids that they are able to spell relief.

The fact that scientific principles led me, a nonseeker, to spiritual insight is appropriate because the latest discoveries in physics and cell research are forging new links between the worlds of Science and Spirit. These realms were split apart in the days of Descartes centuries ago. However, I truly believe that only when Spirit and Science are reunited will we be afforded the means to create a better world.

A Time of Choice

The latest science leads us to a worldview not unlike that held by the earliest civilizations, in which every material object in nature was thought to possess a spirit. The Universe is still thought of as One by the small number of aborigines who survive. Aboriginal cultures do not make the usual distinctions among rocks, air, and humans; all are imbued with spirit, the invisible energy. Doesn't this sound familiar? This is the world of quantum physics, in which matter and energy are completely entangled. And it is the world of Gaia that I spoke of in Chapter 1, a world in which the whole planet is considered to be one living, breathing organism, which needs to be protected from human greed, ignorance, and poor planning.

Never have we needed the insights of such a worldview more. When Science turned away from Spirit, its mission dramatically changed. Instead of trying to understand the "natural order" so that human beings can live in harmony with that order, modern science embarked on a goal of control and domination of nature. The technology that has resulted from pursuing this philosophy has brought human civilization to the brink of spontaneous combustion by disrupting the web of nature. The evolution of our biosphere has been punctuated by five "mass extinctions," including the one that killed the dinosaurs. Each wave of extinction nearly wiped out all life on the planet. As I mentioned in Chapter 1, science now acknowledges that we are "deep" into the sixth mass extinction. Unlike the others caused by galactic forces such as comets, the current extinction is being caused by a force much closer to home—humans. As you sit on your porch and watch the sunset, note its spectacular color. The beauty in the sky reflects the pollution in the air. As the world we know decays, the Earth promises us even greater light shows.

Meanwhile we are leading lives without a moral context. The modern world has shifted from spiritual aspirations to a war for material accumulation. The one with the most toys wins. My favorite image for the scientists and technologists who have championed this path and led us into this spiritless world comes from the Disney movie *Fantasia*. Remember Mickey Mouse as the hapless apprentice to a powerful sorcerer? The sorcerer instructs Mickey to do the chores of the lab while he is away. One of the chores is to fill a giant cistern with water from a nearby well. Mickey, who had been observing the sorcerer's magic, tries to bypass the chore by applying a spell to a broom, which turns it into a water-bucket-carrying lackey.

When Mickey falls asleep, the robotic broom fills and then overfills the cistern, flooding the lab. Upon awakening, Mickey tries to stop the broom. But his knowledge is so limited, he fails and the situation gets even worse. The water takes over, until the sorcerer, who does have the knowledge to quiet the broom, returns and restores balance. Here's how Mickey's predicament is described in the movie: "This piece is a legend about a sorcerer who had an apprentice. He was a bright young lad, very anxious to learn the business. As a matter of fact, he was a little too bright because he had started practicing some of the boss's magic tricks before learning how to control them." Today's very bright scientists are "Mickey Mousing around" with our genes and our environment without understanding how interconnected everything on this planet is—a course of action bound to have tragic results.

How did we get to this point? There was a time when it was necessary for scientists to split from the realm of Spirit, or at least the corruption of Spiritual awareness by the Church. This powerful institution was in the business of suppressing scientific discovery when it was at odds with Church dogma. It was Nicolaus Copernicus, a savvy politician as well as a gifted astronomer, who launched the Spirit/Science split when he released to the public his profound manuscript *De revolutionibus orbium coelestium (On the Revolution of the Heavenly Spheres)*. The 1543 manuscript boldly declared that the sun, not the Earth, was the center of the "Heavenly Spheres." This is obvious today, but in Copernicus' time it was heresy because his new cosmology was at odds with an "infallible" Church truth, which had declared the Earth to be the center of God's firmament. Copernicus believed that the Inquisition would destroy both him and his heretical beliefs, so he prudently waited until he was on his deathbed to publish his work. His concern for his safety was fully justified. Fifty-seven years later Giordano Bruno, a Dominican monk who had the temerity to speak out and defend Copernicus' cosmology, was burned at the stake for this heresy. Copernicus outsmarted the Church—it is hard to torture an intellectual when he is in his grave. Unable to kill the messenger, the Church eventually had to deal with Copernicus' message.

A century later French mathematician and philosopher René Descartes insisted on using scientific methodology to examine the validity of all previously accepted "truths." The invisible forces of the spiritual world clearly didn't lend themselves to such analysis. In the post Reformation era, scientists were encouraged to pursue their studies of the natural world and spiritual "truths" were relegated to the realms of religion and metaphysics. Spirit and other metaphysical concepts were devalued as "unscientific" because their truths could not be assessed by the analytic methods of science. The important "stuff" about life and the Universe became the domain of rational scientists.

If the Spirit/Science split needed any more reinforcement, it got it in 1859 when Darwin's theory of evolution made an instant splash. Darwin's theory spread across the globe like today's Internet rumors. It was readily accepted because its principles dovetailed with people's experiences in breeding pets, farm animals, and plants. Darwinism attributed the origins of humanity to the happenstance of hereditary variations, which meant that there was no need to invoke divine intervention in our lives or our science. Modern scientists were no less awed by the Universe than the cleric/scientists who preceded them, but with Darwin's theory in hand they no longer saw a need to invoke the Hand of God as a grand "designer" of nature's complex order. Preeminent Darwinist Ernst Mayr wrote: "When we ask how this perfection is brought about, we seem to find only arbitrariness, planlessness, randomness, and accident . . . " (Mayr 1976)

While Darwinian theory specifies that the purpose of life's struggles is survival, it does not specify a means that should be used in securing that end. Apparently, "anything goes" in the perceived struggle because the goal is simply survival—by *any* means. Rather than framing the character of our lives by the laws of morality, the neo-Darwinism of Mayr suggests that we live our lives by the law of the jungle. Neo-Darwinism essentially concludes that those who have more deserve it. In the West, we have accepted the inevitability of a civilization that is characterized by the "haves" and the "have-nots." We don't want to deal with the fact that everything in this world has a price. Unfortunately this includes, along with the ailing planet, the homeless, as well as the child laborers who sew our designer jeans . . . *they* are the losers in this struggle.

We Are Made in the Image of the Universe

On that early morning in the Caribbean, I realized that even the "winners" in our Darwinian world are losers because we are one with a bigger Universe/God. On a personal note, I do not perceive of God as a bearded old man on a throne in Heaven. To me, God represents "All That Is," the whole environment comprising the Universe. The cell engages in behavior when its brain, the membrane, responds to environmental signals. In fact, every functional protein in our body is made as a complementary "image" of an environmental signal. If a protein did not have a complementary signal to couple with, it would not function. This means, as I concluded in that "aha!" moment, that every protein in our bodies is a physical/electromagnetic complement to something in the environment. Because we are machines made out of protein, by definition we are made in the image of the environment, that environment being the Universe, or to many, God.

Back to the winners and losers. Because humans evolved as complements of their surrounding environment, if we change the environment too much, we will no longer be complementary to it . . . we won't "fit." At the moment, humans are altering the planet so dramatically that we are threatening our own survival as well as the survival of other, rapidly disappearing organisms. That threat encompasses Hummer drivers and fast food moguls with lots of money, the "winners," along with poverty-stricken laborers, the "losers," in this competition for survival. There are two ways out of this dilemma: to die or mutate. I think you should seriously ponder this as the need to sell Big Macs leads us to decimate the rain forests, as the staggering numbers of gas-guzzling vehicles foul the air, or as petrochemical industries erode the Earth and pollute the water. We were designed by nature to fit an environment but not the environment we are now making.

I learned from cells that we are part of a whole and that we forget this at our peril. But I also recognized that each one of us has a unique, biological identity. Why? What makes each person's cellular community unique? On the surface of our cells is a family of identity receptors, which distinguish one individual from another.

A well-studied subset of these receptors, called self-receptors, or human leukocytic antigens (HLA), are related to the functions of the immune system. If your self-receptors were to be removed, your cells would no longer reflect your identity. These self-receptor-less cells would still be human cells, but without an identity they would simply be generic human cells. Put your personal set of self-receptors back on the cells and they again reflect your identity.

When you donate an organ, the closer your set of self-receptors matches the receptors of the person who is to receive the organ, the less aggressive the rejection reaction launched by the recipient's immune system. For example, let's say that a set of a hundred different self-receptors on the surface of each cell is used to identify you as an individual. You are in need of an organ graft to survive. When my set of one hundred self-receptors is compared to your self-receptors, it turns out that we have only ten receptors that match. I would not be a great organ donor for you. The very dissimilar nature of our self-receptors reveals that our identities are very different. The vast difference in membrane receptors would mobilize your immune system, shifting it into hyper-drive to eliminate the foreign, i.e., not-self, transplanted cells. You would have a greater chance of success if you could find a donor whose self-receptors more closely match the ones on your cells.

In your search for a better donor, however, you will not find a perfect 100 percent match. So far scientists have never found two individuals who are biologically the same. However, it is theoretically possible to create universal donor tissues when you remove the cells' self-receptors, though scientists have yet to carry out such an experiment. In pursuing this proposed experiment, the cells would lose their identity. These self-receptor-less cells would not be rejected. While scientists have focused on the nature of these immune-related receptors, it is important to note that it is not the protein receptors but what activates the receptors that give individuals their identity. Each cell's unique set of identity receptors are located on the membrane's outer surface, where they act as "antennas," downloading complementary environmental signals. These identity receptors *read* a signal of "self," which does not exist within the cell but comes to it from the external environment.

Consider the human body as a television set. You are the image on the screen. But your image did not come from inside the television. Your identity is an environmental broadcast that was received via an antenna. One day you turn on the TV and the picture tube has blown out. Your first reaction would be, "Oh, #*$?!! The television is dead." But did the image die along with the television set? To answer that question you get another television set, plug it in, turn it on, and *tune* it to the station you were watching before the picture tube blew out. This exercise will demonstrate that the broadcast image is still on the air, even though your first television "died." The death of the television as the receiver in no way killed the identity broadcast that comes from the environment.

In this analogy, the physical television is the equivalent of the cell. The TV's antenna, which downloads the broadcast, represents our full set of identifying receptors and the broadcast represents an environmental signal. Because of our preoccupation with the material Newtonian world, we might at first assume that the cell's protein receptors *are* the "self." That would be the equivalent of believing that the TV's antenna is the source of the broadcast. The cell's receptors are not the source of its identity but the vehicle by which the "self" is downloaded from the environment.

When I fully understood this relationship I realized that my identity, my "self," exists in the environment whether my body is here or not. Just as in the TV analogy, if my body dies and in the future a new individual (biological "television set") is born who has the same exact set of identity receptors, that new individual will be downloading "me." I will once again be present in the world. When my physical body dies, the broadcast is still present. My identity is a complex signature contained within the vast information that collectively comprises the environment.

Supporting evidence for my belief that an individual's broadcast is still present even after death comes from transplant patients who report that along with their new organs come behavioral and psychological changes. One conservative, health-conscious New Englander, Claire Sylvia, was astonished when she developed a taste for beer, chicken nuggets, and motorcycles after her heart-lung transplant. Sylvia talked to the donor's family and found she had the heart of an eighteen-year-old motorcycle enthusiast who loved chicken nuggets and beer. In her book called *A Change of Heart,* Sylvia outlines her personal transformational experiences, as well as similar experiences of other patients in her transplant support group. (Sylvia and Novak 1997) Paul P. Pearsall presents a number of other such stories in his book, *The Heart's Code: Tapping the Wisdom and Power of Our Heart Energy.* (Pearsall 1998) The accuracy of memories that accompany these transplants is beyond chance or coincidence. One young girl began having nightmares of being murdered after her heart transplant. Her dreams were so vivid that they led to the capture of the murderer who killed her donor.

One theory about how these new behaviors become implanted into the transplant recipient along with the organ is "cellular memory," i.e., the notion that somehow memories are embedded in cells. You know I have immense respect for the intelligence of single cells, but I have to draw a line here. Yes, cells can "remember" that they are muscle cells or liver cells, but there is a limit to their intelligence. I do not believe cells are physically endowed with perception mechanisms that can distinguish and remember a taste for chicken nuggets!

Psychological and behavioral memory does make sense if we realize that the transplanted organs still bear the original identity receptors of the donor and are apparently still downloading that same environmental information. Even though the body of the person who donated the organs is dead, their broadcast is still on. They are, as I realized in my flash of insight while mulling over the mechanics of the cellular membrane—immortal, as I believe we all are.

Cells and organ transplants offer a model not only for immortality but also for reincarnation. Consider the possibility that an embryo in the future displays the same set of identity receptors that I now possess. That embryo will be tuned into my "self." My identity is back but playing through a different body. Sexism and racism become ridiculous as well as immoral when you realize that your receptors could wind up on a white person, a black person, an Asian, or a male or female. Because the environment represents "All That Is" (God) and our self-receptor antennas download only a narrow band of the whole spectrum, we all represent a small part of the whole . . . a small part of God.

Earth Landers

While the TV analogy is useful, it is not a complete one because a television is only a playback device. In the course of our lives, what we do alters the environment. We change the environment simply by being here. So a more complete way of understanding our relationship to Spirit is to compare a human to the Martian rovers "Spirit" and "Opportunity" or the other NASA landers we have sent to the Moon and Mars. Humans are not yet able to go physically to Mars, but we really want to know what it would be like to land on Mars. So we send up the equivalent of a human explorer. Although the Mars rovers don't physically resemble a human, they have functions of humans. These vehicles have cameras, which are the "eyes" that see the planet. They have vibration detectors, which are "ears" that hear the planet. They have chemical sensors, which "taste" the planet, etc. So the lander is designed with sensors that can experience Mars somewhat as a human would experience it.

But let's look a little more closely at how the Mars rovers work. The rovers have antennas ("receptors") that are tuned to receive information broadcasts by a human being in the form of a NASA controller. The earthbound controller actually sends information that animates the Mariner on Mars. But the information is not a one-way street. The NASA controller also learns from the lander, because the vehicle transmits information about its Mars experiences back to Earth. The NASA controller interprets the information about the lander's experiences and then applies that new awareness to better navigate the Martian terrain.

You and I are like "Earth landers" who receive information from an environmental controller/Spirit. As we live our lives, the experiences of our world are sent back to that controller, our Spirit. So the character of how you live your life influences the character of your "self." This interaction corresponds to the concept of karma. When we understand it, we must take heed of the life we live on this planet because the consequences of our life last longer than our bodies. What we do during our lifetime can come back to haunt us or a future version of ourselves.

In the end, these cellular insights serve to emphasize the wisdom of spiritual teachers throughout the ages. Each of us is a spirit in material form. A powerful image for this spiritual truth is the way light interacts with a prism.

When a beam of white light goes through a prism, the prism's crystalline structure refracts the exiting light so that it appears as a rainbow spectrum. Each color, though a component of the white light, is seen separately because of its unique frequency. If you reverse this process by projecting a rainbow spectrum through the crystal, the individual frequencies will recombine, forming a beam of white light. Think of each human being's identity as an individual color frequency within the rainbow spectrum. If we arbitrarily eliminate a specific frequency, a color, because we don't "like it," and then try to put the remaining frequencies back through the prism, the exiting beam will no longer be white light. By definition, white light is composed of *all* of the frequencies.

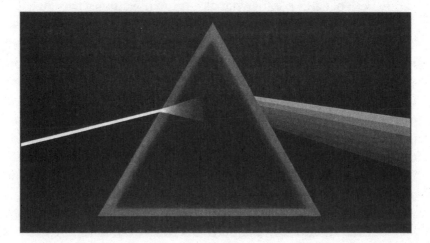

Many spiritual people anticipate the return of White Light to the planet. They imagine that it will come in the form of a unique individual like Buddha, Jesus, or Muhammad. However, from my newly acquired spirituality, I see that White Light will only return to the planet when every human being recognizes every other human being as an individual frequency of the White Light. As long as we keep eliminating or devaluing other human beings we have decided we don't like, i.e., destroying frequencies of the spectrum, we will not be able to experience the White Light. Our job is to protect and nurture each human frequency so that the White Light can return.

Fractal Evolution—A Theory We Can Live With

I've explained why I am now a spiritual scientist. Now I'd like to explain why I am an optimist. The story of evolution is, I believe, a story of repeating patterns. We are at a crisis point, but the planet has been here before. Evolution has been punctuated with upheavals, which virtually wiped out existing species, including the best-known casualties, the dinosaurs. Those upheavals were directly linked to environmental catastrophes just as today's crisis is. As the human population increases, we are competing for space with the other organisms with whom we share the planet. But the good news is that similar pressures in the past have brought into being a new way of living and will do so again. We are concluding one evolutionary cycle and preparing to embark upon another. As this cycle comes to an end, people are becoming understandably apprehensive and alarmed by the failures in the structures that support civilization. I believe, however, that the "dinosaurs" that are currently raping nature will become extinct. The survivors will be those who realize that our thoughtless ways are destructive to the planet and to us.

How can I be so sure? My certitude comes from my study of fractal geometry. Here's a definition of geometry, which will explain why it is important for studying the structure of our biosphere. Geometry is a mathematical assessment of "the way the

different parts of something fit together in relation to each other."
Until 1975, the only geometry available for study was Euclidean,
which was summarized in the thirteen-volume ancient Greek text,
The Elements of Euclid, written around 300 B.C. For spatially ori-
ented students, Euclidian geometry is easy to understand because
it deals with structures like cubes and spheres and cones that can
be mapped on graph paper.

However, Euclidian geometry does not apply to nature. For
example, you cannot map a tree, a cloud, or a mountain using the
mathematical formulas of this geometry. In nature, most organic
and inorganic structures display more irregular and chaotic-
appearing patterns. These natural images can only be created by
using the recently discovered mathematics called fractal geometry.
French mathematician Benoit Mandelbrot launched the field of
fractal mathematics and geometry in 1975. Like quantum physics,
fractal (fractional) geometry forces us to consider those irregular
patterns, a quirkier world of curvy shapes and objects with more
than three dimensions.

The mathematics of fractals is amazingly simple because you
need only one equation, using only simple multiplication and
addition. The same equation is then repeated ad infinitum. For
example, the "Mandelbrot set" is based on the simple formula
of taking a number, multiplying it by itself and then adding the
original number. The result of *that* equation is then used as the
input of the subsequent equation; the result of that equation is then
used as the input for the next equation and so on. The challenge
is that even though each equation follows the same formula, these
equations must be repeated millions of times to actually visualize
a fractal pattern. The manual labor and time needed to complete
millions of equations prevented early mathematicians from recognizing the value of fractal geometry. With the advent of powerful
computers Mandelbrot was able to define this new math.

Inherent in the geometry of fractals is the creation of ever-
repeating, "self-similar" patterns nested within one another. You
can get a rough idea of the repeating shapes by picturing the eter-
nally popular toy, hand-painted Russian nesting dolls. Each smaller
structure is a miniature, but not necessarily an exact version of the

larger form. Fractal geometry emphasizes the relationship between the patterns in a whole structure and the patterns seen in parts of a structure. For example, the pattern of twigs on a branch resembles the pattern of limbs branching off the trunk. The pattern of a major river looks like the patterns of its smaller tributaries. In the human lung, the fractal pattern of branching along the large bronchus repeats in the smaller bronchioles. The arterial and venous blood vessels and the peripheral nervous system also display similar repeating patterns.

Are the repetitive images observed in nature simply coincidence? I believe the answer is definitely "no." To explain why I believe fractal geometry defines the structure of life, let's revisit two points.

First, the story of evolution is, as I've emphasized many times in this book, the story of ascension to higher awareness. Second, in our study of the membrane, we defined the receptor-effector protein complex (IMPs) as the fundamental unit of awareness/ intelligence. Consequently, the more receptor-effector proteins (the olives in our bread-and-butter sandwich model) an organism possesses, the more awareness it can have and the higher it is on the evolutionary ladder.

However, there are physical restrictions for increasing the number of receptor-effector proteins that can be packed into the cell's membrane. The cell membrane's thickness measures seven to eight nanometers, a dimension predetermined by the size of the molecules comprising the phospholipid bilayer. By comparison, the average diameter of the receptor-effector "awareness" proteins is approximately the same as the phospholipid barrier in which they are embedded. Because the membrane's thickness is so tightly defined, you can't cram in lots of IMPs by stacking them on top of one another. You're stuck with a one-protein-thick layer. Consequently, the only option for increasing the number of awareness proteins is to increase the surface area of the membrane.

Let's go back to our membrane "sandwich" model. More olives mean more awareness—the more olives you can layer in the sandwich, the smarter the sandwich. Which has more intelligence capacity, a slice of cocktail rye or a large slab of sourdough? The answer is simple: the larger the surface area of the bread, the greater the number of olives that can fit into the sandwich. Relating this analogy to biological awareness, the more membrane surface area the cell has, the more protein "olives" it can manage. Evolution, the expansion of awareness, can then be physically defined by the increase of membrane surface area. Mathematical studies have found that fractal geometry is the best way to get the most surface area (membrane) within a three-dimensional space (cell). Therefore, evolution becomes a fractal affair. Repeating patterns in nature become a necessity, not a coincidence, of "fractal" evolution.

My point is not to get caught up in the mathematical details of the modeling. There are repetitive fractal patterns in nature and in evolution as well. The strikingly beautiful, computer-generated pictures that illustrate fractal patterns should remind us that, despite our modern angst and the seeming chaos of our world, there is order in nature, and there is nothing truly new under the sun. Evolution's repetitive, fractal patterns allow us to predict that humans will figure out how to expand their consciousness in order to climb another rung of the evolutionary ladder. The exciting, esoteric world of fractal geometry provides a mathematical model that suggests that the "arbitrariness, planlessness, randomness, and accident" that Mayr wrote about is an outmoded concept. In fact, I believe it is an idea that does not serve humanity and should, as rapidly as possible, go the way of the pre-Copernican Earth-centered universe.

Once we realize that there are repeating, ordered patterns in nature and evolution, the lives of cells, which inspired this book and the changes in my life, become even more instructive. For billions of years, cellular living systems have been carrying out an effective peace plan that enables them to enhance their survival as well as the survival of the other organisms in the biosphere. Imagine a population of trillions of individuals living under one roof in a state of perpetual happiness. Such a community exists—it is called the healthy human body. Clearly cellular communities work better than human communities—there are no left-out, "homeless" cells in our bodies. Unless of course, our cellular communities are in profound disharmony causing some cells to withdraw from cooperating with the community. Cancers essentially represent homeless, jobless cells that are living off the other cells in the community.

If humans were to model the lifestyle displayed by healthy communities of cells, our societies and our planet would be more peaceful and vital. Creating such a peaceful community is a challenge because every person perceives the world differently. So essentially, there are more than six billion human versions of reality on this planet, each perceiving its own truth. As the population grows, they are bumping up against each other.

Cells faced a similar challenge in early evolution as described in Chapter 1, but the point bears repeating. Shortly after the Earth was formed, single-celled organisms rapidly evolved. Thousands of variations of unicellular bacteria, algae, yeast, and protozoa, each with varying levels of awareness, appeared over the next three-and-a-half billion years. It is probable that, like us, those single-celled organisms began to multiply seemingly out of control and to overpopulate their environment. They began to bump up against one another and wonder, *Will there be enough for me?* It must have been scary for them, too. With that new, enforced closeness and the consequent change in their environment, they searched for an effective response to their pressures. Those pressures led to a new and glorious era in evolution, in which single cells joined together in altruistic multicellular communities. The end result was humans, at or near the top of the evolutionary ladder.

Similarly, I believe that the stresses of the increasing human population will be responsible for pushing us up another rung on the evolutionary ladder. We will, I believe, come together in a *global* community. The members of that enlightened community will recognize that we are made in the image of our environment, i.e., that we are divine, and that we have to operate, not in a survival of the fittest manner, but in a way that supports everyone and everything on this planet.

Survival of the Most Loving

You may agree that Rumi's words on the power of love are noble ones, but you may not believe that they fit these troubled times, when survival of the fittest may seem more appropriate. Isn't

Darwin right that violence is at the core of life? Isn't violence the way of the natural world? What about all those documentaries that show animals stalking animals, animals snaring animals, animals killing animals? Don't humans possess an inborn inclination to violence? The logic goes: animals are violent, humans are animals, and therefore humans are violent.

No! Humans are not "stuck" with an innate, viciously competitive nature any more than we are stuck with genes that make us sick or make us violent. Chimps, who are the closest to humans genetically, offer evidence that violence is not a necessary part of our biology. One species of chimps, the bonobos, create peaceful communities with co-dominant males and females in charge. Unlike other chimps, the community of bonobos operates not with a violence-driven ethic but an ethic that can be described as "make love, not war." When the chimps in this society become agitated, they don't engage in bloody fights; they diffuse their divisive energy by having sex.

Recent research by Stanford University biologists Robert M. Sapolsky and Lisa J. Share has found that even wild baboons, among the most aggressive animals on this planet, are not genetically mandated to be violent. (Sapolsky and Share 2004) In one well-studied baboon troop, the aggressive males died out from contaminated meat they foraged from a tourist garbage pit. In the wake of their deaths the social structure of the troop was reinvented. Research suggests that females helped steer the remaining, less aggressive males into more cooperative behaviors, which led to a uniquely peaceful community. In an editorial in *Public Library of Science Biology*, where the Stanford research was published, chimp researcher, Frans B. M. de Waal of Emory University, wrote: "even the fiercest primates do not forever need to stay this way." (de Waal 2004)

In addition, no matter how many *National Geographic* specials you've watched, there is no dog-eat-dog imperative for humans. We are at the *top* of the predator/prey food chain. Our survival is dependent on eating organisms lower in the hierarchy, but we are not subject to being eaten by organisms higher in the chain. Without natural predators, humans are spared from becoming "prey" and from all the violence that the term implies.

That does not mean that humans are outside the laws of nature, of course, for eventually, we too shall be eaten. We are mortal, and following our demise, one would hope after a long and violence-free life, our corporeal remains will be consumed and recycled back to the environment. Like a snake turning on itself, humans at the top of the food chain will eventually be devoured by organisms that are the lowest in the chain, the bacteria.

But before that snake turns, we may not live a violence-free life. Despite our lofty position on the food chain, we are our own worst enemy. More than any other animal, we turn on ourselves. Lower-level animals sometimes turn on themselves, but most aggressive encounters among members of the same species are limited to threatening postures, sounds, and scents, not death. And in social populations other than humans, the primary cause of intraspecies violence is either the acquisition of air, water, and food required for survival or the selection of mates for propagation.

In contrast, the violence among humans that is directly linked to securing sustenance or in the process of mate selection is quite minimal. Human violence is more often associated with the acquisition of material possessions beyond what is necessary for sustenance or the distribution and purchase of drugs to escape the nightmare world we have created or child and spousal abuse passed down generation after generation. Perhaps the most wide-spread and insidious form of human violence is ideological control. Throughout history, religious movements and governments have repeatedly prodded their constituents into aggression and violence to deal with dissenters and nonbelievers.

Most human violence is neither necessary nor is it an inherent, genetic, "animal" survival skill. We have the ability, and I believe an evolutionary mandate, to stop violence. The best way to stop it is to realize, as I emphasized in the last chapter of this book, that we are spiritual beings who need love as much as we need food. But we won't get to the next evolutionary step by just thinking about it just as we can't change our children's and our lives simply by reading books. Join communities of like-minded people who are working toward advancing human civilization by realizing that

Survival of the Most Loving is the only ethic that will ensure not only a healthy personal life but also a healthy planet.

Remember those underprepared, underappreciated Caribbean students who banded together, like the cells they studied in their histology course, to form a community of successful students? Use them as role models, and you will help ensure a Hollywood ending not just for individuals mired in self-sabotaging beliefs but also for this planet. Use the intelligence of cells to propel humanity one more rung up the evolutionary ladder where the most loving do more than just survive, they thrive.

✳ ✳ ✳

A decade ago, I faced a challenging dilemma when I prepared the manuscript for the first edition of this book. While my mind was very sure of the science regarding the profound role of epigenetics and the cell membrane in controlling behavior and health, my heart wanted to extend these findings to include the concept of an eternal soul or consciousness. My concern though was that the inclusion of spiritual content might undermine or devalue the contribution provided by the book's hard science of cell biology. After all, conventional science was still aggravating itself over the use of the term "mind," so surely they would become apoplectic over the notion of "spirit."

That science and religion should never overlap was the famous suggestion offered in the book *Rock of Ages* by famed evolution scientist Stephen Jay Gould. (Gould 2002) He supported an idea called NOMA, which is an acronym for Non-Overlapping Magisteria. Yes, I too had to look that word up. Magisteria means "domains having great authority." In this case, Gould is referring to the domains of religion and science.

Unlike many scientists who decry "irrational" spirituality and spiritualists who decry solely "rational" scientists, Gould argued that individuals and cultures should cultivate both a life of the spirit and a life of rational inquiry in order to experience the fullness of being human. But he still advocated that science and religion remain two separate cultures, each one minding its own business and going its own way with its own rules, even though both share the goal of creating a coherent understanding of life. I decided to ignore Gould's advice and delve into the realm of Spirit in the first edition because I had changed—I was no longer solely a hyper-rational, religion-phobic scientist.

Now in this anniversary edition of *The Biology of Belief*, I think it's a good idea to ask if holding a belief about the existence of an afterlife influences our biology and behavior. After all, the idea of a nonphysical afterlife goes as far back as 50,000 years ago—Neanderthals buried their dead in graves, some containing tools and other artifacts interned along with their bodies—so the belief that the end of physical life does not end "life" must have already been instilled at that time. (Rendu, et al, 2014) And the answer to the question about whether a belief in an afterlife influences our biology and behavior is a YES.

Since the first edition of this book, scientific data have been accumulating that show that a person's belief in religion or spirituality has a significant impact on their health and vitality. In 2005 when this book was first released, there were about 800 scientific articles published per year on the impact of religion or spirituality on health. Now ten years on, there are 5,000 articles per year written on that topic and they reveal that medical and psychiatric patients commonly resort to religious and spiritual belief practices to cope with illness and other stressful life changes.

Dr. Harold G. Koenig, Professor of Medicine at Duke University, reviewed over 600 of these research studies and concluded that people who hold more spiritual beliefs fare significantly better in mental health and adapt more quickly to health problems than those who are less spiritual. The benefits to mental health and well-being provided by spirituality have physiological consequences that impact physical health, reduce the risk of disease, and influence the healing outcomes of treatment. Spiritual beliefs have a direct, positive influence on the activity of the immune and endocrine systems that are critical for health maintenance and disease prevention. Spiritual patients exhibit significantly better indicators of immune functions, such as higher white blood cell counts and antibody levels and experience significantly lower infection rates. They also exhibit lower levels of adrenal stress hormones, such as cortisol and epinephrine (secretions that directly repress the activity of the immune system) than nonspiritual patients. (Koenig 2012)

The health-promoting effects of spiritual consciousness lie at least in part on the fact that spiritual consciousness offers hope and provides an alternative to our innate fear of death. Most children first experience death while very young when they lose their pet goldfish, parakeet, cat, or dog. So as children we learn from these death experiences that these friends are gone—forever. As sad as that it is, it is not as terrifying as when our consciousness connects the fact that the parents we depend on face the same future. Death becomes most terrifying when we finally realize that we too are mortal.

That's because built into every organism, from bacteria up, is a fundamental behavior mechanism known as the biological imperative, a system of programmed behaviors that are responsible for the "will to survive." This imperative automatically engages behaviors needed to sustain an organism's life, such as breathing air, drinking water, finding food, and protecting oneself from threats. Animals with no awareness of their own mortality engage the imperative's protection responses only in the face of an imminent, life-threatening experience, e.g., the presence of that nasty, old saber-toothed tiger. However, once a human becomes aware of his or her own mortality, the imperative mechanism runs on full-time alert, analyzing the threat potential of every move and intention.

The biological imperative's "will to survive" behaviors are managed by the subconscious mind. As a result, we fortunately do not have to be *consciously* burdened with chronic fears over the potential consequences of our actions. In fact, when an individual's conscious mind does become preoccupied with persistent fears of dying, those fears can lead to death anxiety, a behavioral dysfunction in which people succumb to a chronic feeling of dread, apprehension, or anxiety over the thought of ceasing to "be."

The conscious mind might be blissfully unaware of the subconscious mind's hyper-vigilant activities, but the body's cells, tissues, and organs are unfortunately influenced by the imperative's sustained release of stress hormones. But an individual who truly owns his or her spiritual nature is no longer burdened by the fear of death, a consciousness that creates an unperceived weight on our lives. Personally, I only became aware of that burden when I felt an immediate and unusual lightness the instant my mind made the connection between my cell's self-receptors and my spiritual source. Since then, I have heard the same description of a physical release by many others whose life experiences have precipitated an "instant" spiritual transformation. In every case, a loss of the fear of death leads to a lightening of a heretofore-unrecognized weight and an increase in energy and personal empowerment. It is important to note that the health benefits and ease of life that accompany the belief in a spiritual afterlife work whether there is any reality to that awareness or not. The health impact of spiritual consciousness is another example of the power of belief.

Before we consider evidence for the existence of an afterlife, I have to first mention that modern physics even questions whether death exists at all. A fundamental characteristic of quantum physics is Heisenberg's principle of uncertainty, which stipulates that certain observations cannot be predicted. Instead, they express a range of possibilities, each based on a different probability. A conventional way to account for this multiplicity of outcomes is the "many-worlds" explanation, which states that each of the possible observations corresponds to a different universe; in other words, there is a simultaneous existence of many worlds, called the "multiverse."

A new scientific theory, *biocentrism*, further refines this notion by suggesting that there are an infinite number of universes and everything that could possibly happen occurs in some universe. Theoretically, death cannot exist in any real sense because all possible universes exist simultaneously, regardless of what happens in any of them. Although we shed our bodies, the alive feeling expressed by Descartes' "I am" in his famous quote, "I think, therefore I am," reflects the energy swirling in and around our brains. One of the absolutes of science is that energy can neither be created nor destroyed; consequently, the energy associated with our identity doesn't disappear at death. (Lanza 2009)

Does this mean that our unique energy profile transcends this one world we're aware of and transfers to other worlds? His Holiness of Space-Time and my personal hero, Albert Einstein, wrote the following to a family member of his recently deceased friend Michele Besso: "Now Besso has departed from this strange world a little ahead of me. That means nothing. People like us . . . know that the distinction between past, present, and future is only a stubbornly persistent illusion." Einstein is referring to our illusion of time, a concept he elaborated on in his theory of relativity, which holds that there is no single, special "present" moment and that all moments in time are equally real. If there's no distinction between past, present, and future, then, by definition, death is an illusion. For Einstein, immortality doesn't mean a perpetual existence in time without end, but instead, existence resides outside of time altogether. (Hoffman and Dukas 1972)

That's also the case for those who believe they have experienced past lives. History is replete with anecdotal stories, many of them from remarkable discoveries of individuals who had factual knowledge of people who lived before them in places they had never been. Perhaps one of the most convincing and documented cases of a past life reality is the story of James Leininger published in a book, *Soul Survivor*, written by his parents. (Leininger 2009)

Three weeks after James Leininger's second birthday, he began to experience the same terrifying nightmare, night after night. In his sleep James would shout out recurring phrases such as, "Plane on fire! Little man can't get out!" Out of concern for their son's welfare, Bruce and Andrea Leininger pieced together what their son was communicating and eventually discovered that he was reliving the past life of World War II fighter pilot James Huston. Huston was stationed on the aircraft carrier U.S.S. *Natoma Bay* and died after being shot down in a battle over the Sea of Japan. When young James was taken to a reunion of *Natoma Bay* veterans he recognized many by name. When the Leiningers found out that one of James Huston's siblings was still alive, they contacted her. Through their conversations, James was able to accurately recall Huston family history that he experienced while in his former life as James Huston. *Soul Survivor* is a riveting true story of how the Leiningers' belief system, which did not include a belief in reincarnation, was shaken to the core, as they came to recognize the fact that their little boy, against all odds and in the face of true skeptics, including themselves, nevertheless harbored the soul of a man who had died long ago.

There are thousands of reports of people describing similar life-after-death, nonphysical realities, especially associated with near death experiences (NDEs). NDE events occur when an individual loses consciousness, usually as a result of life-threatening situations, and then have what they refer to as an out-of-body experience. A common scenario is the sense of "going to the Light," which is associated with feelings of overwhelming love and peace. While in the Light, people who experience NDEs often report that they communicate with deceased family members and friends who have already "crossed over."

Like James Leininger's parents, Dr. Eben Alexander was an NDE skeptic. As a highly successful neurosurgeon, Eben had heard many NDE stories from his patients, and like his colleagues, he considered his patients' reports hallucinations. Though he acknowledged that NDEs feel real to those people experiencing them, he believed that in truth, they were simply fantasies produced by brains under extreme stress. What people of faith called the "soul," Dr. Alexander attributed to the activity of brain chemistry.

All of Dr. Alexander's strongly held beliefs about NDE hallucinations radically changed when he suffered his own neurological crisis. Dr. Alexander's brain became infected with a dangerous strain of gram-negative *E. coli* bacteria, an extremely rare and normally lethal infection that eroded his neocortex, the part of the brain that controls thought and emotion. Neocortex activities involve higher brain functions such as sensory perception, motor control over muscles, spatial reasoning, language, and conscious thought. Damage to the neocortex often results in semantic dementia, which is the loss of memory of factual information (in other words, permanent amnesia). The aggressive bacteria were turning Dr. Alexander's brain cortex into a pile of pus. As a result, his conscious processing completely shut down and he spent seven days in a deep coma. (Alexander 2012)

Dr. Alexander's recovery is by all accounts a medical miracle. But the real miracle of his story is that while his body lay in coma, he journeyed beyond this world and encountered an angelic being who guided him into the deepest realms of the supernatural. What makes Dr. Alexander's NDE so unique is that his vivid out-of-body experience occurred while his higher functioning neocortex had turned to mush and his brain was off-line. In other reported NDE incidences, the "travelers" acknowledged that they had retained knowledge of who they were and memories of all the entities that they had encountered. In contrast, Dr. Alexander had no idea of who (or even what) he was beyond being a conscious observer, a reflection of the fact that his brain had lost all memory.

When he awoke from his coma, Dr. Alexander had no recollection of who he was or any other remembrances of his life. Over a period of time, Dr. Alexander's life history was slowly and completely restored piece by piece. This happened in spite of the fact that if memories are stored in the brain's structure, the infection and death of massive numbers of neocortex neurons should have permanently destroyed Dr. Alexander's memory of his identity and previous history. Instead, the return of Dr. Alexander's memory after the regeneration of his cortical brain tissue suggests that, in the manner that programs are received by radios, memories are not built into the brain's structure, but are "downloaded" by a functioning brain. These observations corroborate the conclusions of Dr. Lorber, reported in Chapter 7, that even though hydrocephalus may destroy most of the brain's cerebral cortex, patients can be highly intelligent and still live normal lives.

Perhaps the most empowering of all NDEs is the amazing story of Anita Moorjani and her husband, Danny, reported in her book *Dying to Be Me*. (Moorjani 2012) For me, Anita's life and cancer journey makes her the poster child for *The Biology of Belief*. After four years of battling an aggressive cancer, Anita's body had reached a point of no return. In what was thought to be her final week of life, Anita's systems began to shut down. While on life support, her body began to absorb her tissues—she became so emaciated that many of her cancer growths could be observed as protruding lumps in her skin. When she finally slipped into a coma, her attending oncologist warned the family that Anita would likely not come back into consciousness, and they should prepare for the worst.

While in her coma, Anita left her body and in the process her conscious awareness started to profoundly expand. At first her attention focused on what was happening in her immediate environment in the intensive care wing of her Hong Kong hospital. While in a coma and unconscious, Anita writes that she was aware not only of conversations around her bedside, but as her awareness began to float above her body, she accurately heard conversations among hospital staff in other parts of the building. Her consciousness rose higher and higher and soon she vividly observed her brother, thousands of miles away in India, when he first responded to the news of her coma and impending death.

As with other NDE reports, Anita found herself in a nonphysical reality where she felt profound love, health, and peace, free of her body's painful physical demise. In this otherworldly environment, Anita met her deceased and much beloved father, as well as other family members and acquaintances who had passed on. Through her communication with her father and others and a review of her life's story, Anita came to realize that her cancer was directly tied to life experiences that were traumatic because they conflicted with the cultural programming she had received in her formative years. She clearly perceived how her fears and worries about diverging from her cultural programming were responsible for her severe illness.

As with many NDE reports, Anita was given a choice to return to her body or stay in the loving and serene environment she was experiencing. While Anita would have opted to not return to her disease-ravaged body, she realized her death would be a severe blow to the life and health of her beloved husband, Danny. Danny, who had not let go of her hand since she went into a coma, had left his job and stayed by her side, taking care of her needs for several years during her illness.

When she did return to her body, Anita brought with her all the lessons she had learned in her otherworldly travels. The result was that Anita woke up and shocked the medical community with the immediate recovery of her failed vital bodily functions. Within two weeks of coming out of the coma, a bone marrow biopsy revealed no trace of the cancer that had almost killed her. I found the following passage in her book about her doctor's response quite amusing:

> "We have the results of the bone-marrow biopsy, but it's a little disturbing."

> For the first time in days, I felt some anxiety. "Why? What's the problem?"

> My family members were in the hospital room with me, and all of them looked worried.

"We can't find the cancer in your bone-marrow biopsy," he said.

"So how is that a problem?" Danny asked. "Doesn't that just mean she doesn't have cancer in her bone marrow?"

"No, that's not possible," the doctor said. "She definitely has cancer in her body—it can't just disappear so quickly like that. We simply have to find it."

While doctors were persistent in their quest to find her "missing" cancer, five weeks after coming out of her coma, Anita was released from the hospital cancer free! Anita recognized the role belief played in her illness, writing, "I felt a level of victory. I'd so completely overcome my fear of everything—from dying to cancer to chemotherapy—that this proved to me that it had been the fear destroying me." Yes, as I've emphasized over and over in this book, fear kills.

Today, Anita and Danny travel the world, sharing stories of how the opposite of fear, i.e., Love, conquered terminal cancer and of the special insights Anita gained from her otherworldly travels. I highly recommend Anita's book for everyone, and especially those facing life-threatening illness, not only for the medical implications of her radical change in belief, but also for the amazing wisdom she acquired about the nature of our lives on this planet and the fate of our souls when we leave our mortal remains.

Crisis precipitates evolution. From the state of our planet, it is apparent that in order to survive the current global crises as individuals and as a species, we must evolve. As the life experiences of Anita Moorjani and thousands of others demonstrate, the insights and empowerment offered in *The Biology of Belief* can profoundly aid that evolution and help heal the world.

ADDENDUM

The science revealed in this book defines how *beliefs* control behavior and gene activity and, consequently, the unfolding of our lives. The chapter on Conscious Parenting describes how most of us unavoidably acquired limiting or self-sabotaging *beliefs* that were downloaded into our subconscious minds when we were children.

A child's perceptions of the world are directly downloaded into the subconscious during the first six years of life, without discrimination and without filters of the analytical self-conscious mind that is not fully operational during this time. Consequently, our fundamental perceptions about life and our role in it are learned without our having the capacity to choose or reject those beliefs. Since the subconscious mind controls about 95 percent of our behavior, other people essentially program our lives.

The Jesuits were aware of this programmable state and proudly boasted, "Give me the child until he is seven, and I will give you the man." They were aware that the child's theta trance state facilitated a direct implanting of Church dogma into the subconscious mind. Once programmed, that information would inevitably influence 95 percent of that individual's behavior for the rest of his or her life.

As I mentioned earlier, there are a variety of exciting new techniques that exploit the latest mind-body research to quickly access and reprogram those subconscious programs. On the Resource page of my website, www.brucelipton.com, I provide a partial listing of available belief change modalities. There is no one source that works for all people. Based on the influence of the placebo and nocebo effects, the reprogramming modality that works best is the one in which you hold the most belief.

I will only delve deeper into one of these energy psychology techniques called PSYCH-K because I have personal experience with it, and I am confident of its integrity, simplicity, and effectiveness.

I met Rob Williams, the originator of PSYCH-K, at a conference in 1990, where we were both presenters. As usual, at the conclusion of my presentation, I told my audience that if they changed their *beliefs* they could change their lives. It was a familiar conclusion with a familiar response from the participants: "Well, Bruce, that's great, but how do we do that?"

In those days I didn't fully realize the crucial role the *subconscious* mind plays in the change process. Instead, I relied mostly on trying to power through negative behavior using positive thinking and willpower. I knew, though, that I had had only limited success in making personal changes in my own life. I also knew that when I offered this solution, the energy in the room dropped like a lead balloon. It seems my sophisticated audiences had, like me, already tried willpower and positive thinking with limited success!

As fate would have it, I returned to my seat, and looked up to see that the next presenter was psychotherapist Rob Williams. Rob's opening remarks quickly had the entire audience on the edge of our seats. In his introduction, Rob stated that PSYCH-K can change long-standing, limiting beliefs in a matter of minutes.

Rob then asked the audience if there was anyone who would like to address an issue that had been troubling them. One woman caught both Rob's and my attention. She raised her hand tentatively, first up, then down, then up again. Her timidity was palpable. When Rob asked what her issue was, her face turned red and her response was inaudible. Rob actually had to leave the podium and confer with her, one-on-one. It was Rob who had to inform the audience that her problem was "speaking in public." Rob returned to the stage and the woman hesitantly followed. Rob asked her to tell the audience of nearly a hundred people a bit about her fear. Again she could hardly speak.

Rob worked with the woman for about ten minutes, using one of the PSYCH-K change techniques. Then he again asked the woman to tell the audience how she felt about speaking to them.

The change was astonishing. Not only was she visibly more relaxed, she began talking to the audience in an excited, yet confident voice. The eyes of the conference attendees became like saucers and their mouths dropped as this woman took over the stage for the next five minutes. The woman got so carried away that Rob had to ask her to stop speaking and take her seat so he could finish his presentation!

Because this woman was a regular attendee at an annual conference, and I was a frequent presenter, I was able to witness her amazing transformation over the next several years. She not only had gotten over her fear of public speaking, but even went on to organize Toastmasters in her community. Eventually, she became an award-winning public speaker! This woman's life was truly transformed in just a few minutes. In the fifteen years since I witnessed that woman's quick transformation, I have seen other people rapidly improve their self-esteem and change their relationships, their finances, and their health using PSYCH-K.

The PSYCH-K process is simple, direct, and verifiable. It utilizes the mind/body interface of muscle testing (kinesiology) that I first discovered in a student-chiropractor's makeshift office in the Caribbean, to access the self-limiting "files" of the subconscious mind. It also makes use of left brain/right brain integration techniques to effect swift and long-lasting changes. In addition, PSYCH-K integrates Spirit into the change process, just as I have integrated Spirit into my understanding of Science. Using muscle testing, PSYCH-K accesses what Rob calls the "superconscious" mind to make sure that the person's stated goals are safe and appropriate. These built-in safeguards allow this system of personal change to be taught to anyone who is interested in taking charge of their lives by moving out of fear and into love.

✳ ✳ ✳

Since 2005, when this book was first published, new studies reveal that the self-empowerment offered by PSYCH-K balances is more than just a subjective, anecdotal experience. Research by neuroscientist Jeffrey L. Fannin, Ph.D., an expert in computerized brain mapping, shows that a PSYCH-K balance produces an objective and radical change in brain EEG activity, leading to a balanced brain wave energy pattern referred to as the "whole-brain state." This is a state of coherency in the brain marked by a bilateral, symmetrical brain wave pattern that allows for maximum communication and data flow between the left and right hemispheres of the brain. Fannin reports that the PSYCH-K belief modification process enhances our functionality by optimizing belief systems and brain function. (Fannin and Williams 2012)

I use PSYCH-K in my own life. PSYCH-K has helped me undo my self-limiting beliefs, including one about not being able to finish my book. The fact that you are holding this book is one indication of the power of PSYCH-K! While this book is about the New Biology, I believe that PSYCH-K represents an important step toward the New Psychology for the twenty-first century and beyond. You can find more information about PSYCH-K at Rob's website: **www.psych-k.com.**

For scientific updates and more information, visit:
www.brucelipton.com

- Free downloadable articles and references
- Books, videotapes, and DVDs
- Seminar and workshop schedules
- Links to other valuable websites

Enjoy the full impact of Dr. Lipton's startlingly clear science and dynamic presentation style captured live on video. These masterful works from an award-winning teacher make science simple and our evolution as humans more hopeful.

Watch as the concepts introduced in *The Biology of Belief: Unleashing the Power of Consciousness, Matter, and Miracles* are brought to life in three remarkable presentations. Own the video library that brings Science and Spirit together as you have never before seen.

Visit www.brucelipton.com or call toll free 800-550-5571.

ACKNOWLEDGMENTS

Much has transpired between my scientific inspiration and the creation of this book. During this time of great personal transformation, I was blessed and guided by both spiritual and incarnate muses—the inspiring spirits of the arts. I am particularly indebted to the following muses who have helped make this book a reality.

The Muses of Science: I am indebted to the *spirits* of science, for I am fully aware that *forces* outside of myself have guided me in bringing this message to the world. Special blessings to my heroes, Jean-Baptiste de Monet Lamarck and Albert Einstein, for their world-changing spiritual and scientific contributions.

The Muses of Literature: The intention to write a book on the New Biology was spawned in 1985, though it was not until Patricia A. King came into my life in 2003 that this book could come into reality. Patricia is a Bay Area freelance writer and former *Newsweek* reporter who worked for a decade as the magazine's San Francisco Bureau Chief. I will never forget our first meeting wherein I overwhelmed her with a lengthy New Science tutorial and then burdened her with a truckload of aborted manuscripts, sheaves of innumerable articles I had written, boxes overflowing with videotaped lectures, and stacks of scientific reprints.

Only as she was driving away did I realize the monumental nature of the task I was asking of her. Without formal training in cell biology and physics, Patricia accomplished miracles in downloading and understanding the New Science. In a very short time, she not only learned the New Biology, she was even able to expand on its topics. Her amazing skills at integrating, editing, and synthesizing information are responsible for the clarity of this book.

Patricia works on book projects and newspaper and magazine stories that focus on health issues, especially mind-body medicine and the role stress plays in disease. Her work has appeared in publications such as the *Los Angeles Times,* Southwest Airlines' *Spirit* magazine, and *Common Ground* magazine. A native of Boston, King lives in Marin with her husband, Harold, and their daughter, Anna. I am in deep appreciation and gratitude to Patricia for her efforts and look forward to the opportunity of writing another book with her in the future.

The Muses of the Arts: In 1980 I left academia and went "on the road" presenting a touring light show called *The Laser Symphony.* The heart and brains of our spectacular laser production was Robert Mueller, a visionary artist and computer graphics genius. Wise beyond his teenage years, Bob drank in the New Science I was working on, first as a student and later as my "spiritual son." Years ago he offered, and I accepted, his bid to create a cover for the book whenever it would appear.

Bob Mueller is cofounder and creative director of LightSpeed Design, Bellevue, Washington. He and his company have produced award-winning 3-D light and sound shows for science museums and planetariums around the world. Their edutainment show on the fragile ecology of our oceans was an honored presentation, seen by 16,000 viewers daily at the World's Expo held in Lisbon, Portugal (1998). Bob's creative endeavors can be sampled at **www.lightspeeddesign.com**.

Bob's work, inspired by science and the Light, is beautiful and profound. I am honored to have his contribution as the cover art—the image that will introduce this new awareness to the public.

Muses of Music: From the conception of this New Science to the submission of the book, I have been continuously encouraged and energized by the music of Yes and especially the lyrics of their vocalist Jon Anderson. Their music and message reveal an inner knowing and understanding of the new science. The music of Yes speaks to the fact that we are all connected to the Light. Their songs emphasize how our experiences, our beliefs, and our dreams shape our lives and influence those of our children. What takes me pages of text to explain, Yes can say in a few powerful and poignant lines. You guys are great!

Regarding the physical production of this book, I sincerely want to thank the New York publishers who turned down the book proposal. Without you, I was able to create my *own* book—just like I wanted to do. I am indebted to Mountain of Love Productions, Inc. for investing time and resources into publishing the first edition of this book. Toward that end, special appreciation goes out to Dawson Church of Author's Publishing Cooperative. Dawson enabled us to have the best of both worlds, the personal management afforded by self-publishing and the marketing experience of a major publishing house. Thanks to Geralyn Gendreau for her support of this work and bringing it to the attention of Dawson Church. Dear friend and public relations specialist Shelly Keller has given generously of her time in providing professional editing skills.

Thanks to all of the students and attendees of my classes, lectures, and seminars who over the years persistently inquired, "Where's the book???" Okay, okay, here it is! Your continued encouragement is deeply appreciated.

I would like to honor some very important teachers who have provided guidance in my scientific career. First and foremost, my father, Eli, who instilled in me a sense of purpose and, just as important, encouraged me to "think outside of the box." Thanks, Dad.

David Banglesdorf, the elementary school science teacher who introduced me to the world of cells and ignited my passion for science. The brilliant Irwin R. Konigsberg, Ph.D., who took me under his wing and mentored my doctoral education. I shall forever remember our eureka moments and the passion for science we shared.

I am indebted to Professor Theodore Hollis, Ph.D. (Penn State University), and Klaus Bensch, M.D., Chairman of Pathology (Stanford University), the first "real" scientists who understood my heretical ideas. Each of these distinguished researchers encouraged and supported my efforts by providing me with space in their laboratories to investigate the ideas presented in this book.

In 1995 Gerard Clum, D.C., President, Life College of Chiropractic West, invited me to teach Fractal Biology, my very own course on the New Science. I am grateful for Gerry's support, for he introduced me to the life-enhancing worlds of chiropractic and complementary medicine.

At the first public presentation of this material in 1985 I met Lee Pulos, Ph.D., Assistant Professor Emeritus, Department of Psychology at the University of British Columbia. Over the years, Lee has been a great supporter of and contributor to the New Biology presented in this book. My partner and esteemed colleague Rob Williams, M.A., developer of PSYCH-K, contributed to this project by helping bridge the science of cells with the mechanics of human psychology.

Discussions of science and its role in civilization with Curt Rexroth, D.C., a dear friend and wizard of philosophy, have brought great awareness and joy to my life. Collaboration with Theodore Hall, Ph.D., offered wonderful and profound insights correlating the histories of cellular evolution and human civilization.

I sincerely want to thank Gregg Braden for his wonderful scientific insights, his suggestions regarding publishing, and for providing the intriguing subtitle of this book.

Each of the following dear and trusted friends read and critiqued this work. Their contributions were vital in bringing this book to you. I personally want to thank each of them: Terry Bugno, M.D., David Chamberlain, Ph.D., Barbara Findeisen, M.F.T., Shelly Keller, Mary Kovacs, Alan Mande, Nancy Marie, Michael Mendizza, Ted Morrison, Robert and Susan Mueller, Lee Pulos, Ph.D., Curt Rexroth, D.C., Christine Rogers, Will Smith, Diana Sutter, Thomas Verney, M.D., Rob and Lanita Williams, and Donna Wonder.

I am grateful for the love and support offered by my sister, Marsha, and brother, David. I am particularly proud of David for what he jokingly refers to as "breaking the circle of violence" and becoming a great dad to his son, Alex.

Much appreciation goes out to Doug Parks of Spirit 2000, Inc., for his outstanding support of this project. Upon hearing the New Biology, Doug fully dedicated his efforts to getting this message out to the world. Doug has produced video lectures and workshops that have broadened public awareness of this material and have opened the doors to many seeking self-empowerment. Thank you, dear brother.

These acknowledgments would not be complete without a most special thank you to Margaret Horton. Margaret has been the driving force behind the scenes and has empowered the writing and materialization of this book. What ever I write and say, my darling . . . it was done in love for you!

Endnotes

Introduction

Lipton, B. H. (1977a). "A fine structural analysis of normal and modulated cells in myogenic culture." *Developmental Biology* 60: 26-47.

Lipton, B. H. (1977b). "Collagen synthesis by normal and bromodeoxyuridine-treated cells in myogenic culture." *Developmental Biology* 61: 153-165.

Lipton, B. H., K. G. Bensch, et al. (1991). "Microvessel Endothelial Cell Transdifferentiation: Phenotypic Characterization." *Differentiation* 46: 117-133.

Lipton, B. H., K. G. Bensch, et al. (1992). "Histamine-Modulated Transdifferentiation of Dermal Microvascular Endothelial Cells." *Experimental Cell Research* 199: 279-291.

Chapter One

Adams, C. L., M. K. L. Macleod, et al. (2003). "Complete analysis of the B-cell response to a protein antigen, from in vivo germinal centre formation to 3-D modelling of affinity maturation." *Immunology* 108: 274-287.

Ahmed, Nafeez (2014). "Nasa-funded study: industrial civilisation headed for 'irreversible' collapse?" *The Guardian* March 14, 2014.

Balter, M. (2000). "Was Lamarck Just a Little Bit Right?" *Science* 288: 38.

Barnosky, D., E. A. Hadly, et al. (2012). "Approaching a state shift in Earth's biosphere." *Nature* 486: 52-58.

Biello, David (2010). "Genetically Modified Crop on the Loose and Evolving in U.S. Midwest." *Scientific American* August 6, 2010.

Blanden, R. V. and E. J. Steele (1998). "A unifying hypothesis for the molecular mechanism of somatic mutation and gene conversion in rearranged immunoglobulin variable genes." *Immunology and Cell Biology* 76(3): 288.

Blaser, Martin J. M.D. (2014). *Missing Microbes: How the Overuse of Antibiotics Is Fueling Our Modern Plagues.* New York, Henry Holt.

Boucher, Y., C. J. Douady, et al. (2003). "Lateral Gene Transfer and the Origins of Prokaryotic Groups." *Annual Review of Genetics* 37: 283-328.

Cairns, J., J. Overbaugh, and S. Miller (1988). "The Origin of Mutants." *Nature* 335: 142-145.

Darwin, Charles (1859) (Originally published by Charles Murray in 1859, London) *The Origin of Species by Means of Natural Selection: or The Preservation of Favoured Races in the Struggle for Life* (Reprinted by Penguin Books, London, 1985).

Desplanque, B., N. Hautekeete, et al. (2002). "Transgenic weed beets: possible, probable, avoidable?" *Journal of Applied Ecology* 39(4): 561-571.

Diaz, M. and P. Casali (2002). "Somatic immunoglobulin hypermutation." *Current Opinion in Immunology* 14: 235-240.

Dirzo, R., H. S. Young, et al. (2014). "Defaunation in the Anthropocene." *Science* 345: 401-406.

Dutta, C. and A. Pan (2002). "Horizontal gene transfer and bacterial diversity." *Journal of Biosciences* (Bangalore) 27 (1 Supplement 1): 27-33.

Gearhart, P. J. (2002). "The roots of antibody diversity." *Nature* 419: 29-31.

Gogarten, J. P. (2003). "Gene Transfer: Gene Swapping Craze Reaches Eukaryotes." *Current Biology* 13: R53-R54.

Haygood, R., A. R. Ives, et al. (2003). "Consequences of recurrent gene flow from crops to wild relatives." *Proceedings of the Royal Society of London,* Series B: Biological Sciences 270 (1527): 1879-1886.

Heritage, J. (2004). "The fate of transgenes in the human gut." *Nature Biotechnology* 22(2): 170+.

Jordanova, L. J. (1984). *Lamarck.* Oxford, Oxford University Press.

Lamarck, J.-B. de M., Chevalier de (1809). *Philosophie zoologique, ou exposition des considerations relativesà l'histoire naturelle des animaux.* Paris, Libraire.

Lamarck, J.-B. de M., Chevalier de (1914). *Zoological Philosophy: an exposition with regard to the natural history of animals.* London, Macmillan.

Lamarck, J.-B. de M., Chevalier de (1963). *Zoological philosophy* (facsimile of 1914 edition). New York, Hafner Publishing Co.

Lenton, T. M. (1998). "Gaia and natural selection." *Nature* 394: 439-447.

Li, Y., H. Li, et al. (2003). "X-ray snapshots of the maturation of an antibody response to a protein antigen." *Nature Structural Biology* 10(6).

Lovell, J. (2004). *Fresh Studies Support New Mass Extinction Theory.* Reuters. London.

Mayr, E. (1976). *Evolution and the Diversity of Life: selected essays.* Cambridge, Mass., The Belknap Press of Harvard University Press.

Milius, S. (2003). "When Genes Escape: Does it matter to crops and weeds?" *Science News* 164: 232+.

Morris, Kevin (2012). "Invited Editorial: Lamarck and the Missing Lnc." *The Scientist* 26.

Netherwood, T., S. M. Martín-Orúe, et al. (2004). "Assessing the survival of transgenic plant DNA in the human gastrointestinal tract." *Nature Biotechnology* 22(2): 204+.

Nitz, N., C. Gomes, et al. (2004). "Heritable Integration of kDNA Minicircle Sequences from Trypanosoma cruzi into the Avian Genome: Insights into Human Chagas Disease." *Cell* 118: 175-186.

Nowak, Martin (2012). "Why We Help: Far from being a nagging exception to the rule of evolution, cooperation has been one of its primary architects." *Scientific American* (July 2012): 34-39.

Pennisi, E. (2001). "Sequences Reveal Borrowed Genes." *Science* 294: 1634-1635.

Pennisi, E. (2004) "Researchers Trade Insights About Gene Swapping." *Science* 305: 334-335.

Rogers, Kara (2009). "The Rebirth of Lamarckism (The Rise of Epigenetics)." *Encyclopedia Britannica Online Encyclopedia.*

Ruby, E., B. Henderson, et al. (2004). "We Get By with a Little Help from Our (Little) Friends." *Science* 303: 1305-1307.

Ryan, F. (2002). *Darwin's Blind Spot: Evolution beyond natural selection.* New York, Houghton Mifflin.

Saey, Tina Hesman (2013A). "Year in Review: Your body is mostly microbes." *Science News* 184 (Dec. 28, 2013A).

Saey, Tina Hesman (2013B). "People's genes welcome their microbes: In mice and humans, genetic variants seem to control the bacterial mix on and in bodies." *Science News* 184 (November 30, 2013B).

Spencer, L. J. and A. A. Snow (2001). "Fecundity of transgenic wild-crop hybrids of Cucurbita pepo (Cucurbitaceae): implications for crop-to-wild gene flow." *Heredity* 86: 694-702.

Steele, E. J., R. A. Lindley, et al. (1998). *Lamarck's Signature: how retrogenes are changing Darwin's natural selection paradigm.* St Leonards NSW Australia, Allen & Unwin.

Stevens, C. J., N. B. Dise, et al. (2004). "Impact of Nitrogen Deposition on the Species Richness of Grasslands." *Science* 303: 1876-1879.

Thomas, J. A., M. G. Telfer, et al. (2004). "Comparative Losses of British Butterflies, Birds, and Plants and the Global Extinction Crisis." *Science* 303: 1879+.

Waddington, C. H. (1975). *The Evolution of an Evolutionist.* Cornell, Ithaca, New York.

Watrud, L. S., E. H. Lee, et al. (2004). "Evidence for landscape-level, pollen-mediated gene flow from genetically modified creeping bentgrass with CP4 EPSPS as a marker." *Proc. National Academy of Sciences* 101(40):14533-14538.

Whittaker, R. J., M. B. Bush, and K. Richards (1989). "Plant Recolonization and Vegetation Succession on the Krakatau Islands, Indonesia." *Ecological Monographs* 59(2): 59-123.

Wu, X., J. Feng, et al. (2003). "Immunoglobulin Somatic Hypermutation: Double-Strand DNA Breaks, AIDs and Error-Prone DNA Repair." *Journal of Clinical Immunology* 23(4).

Chapter Two

Avery, O. T., C. M. MacLeod, et al. (1944). "Studies on the chemical nature of the substance inducing transformation of pneumococcal types. Induction of transformation by a deoxyribonucleic acid fraction isolated from Pneumococcus Type III." *Journal of Experimental Medicine* 79: 137-158.

Baltimore, D. (2001). "Our genome unveiled." *Nature* 409: 814-816.

Baylin, S. B. (1997). "DNA METHYLATION: Tying It All Together: Epigenetics, Genetics, Cell Cycle, and Cancer." *Science* 277(5334): 1948-1949.

Blackburn, E. and E. Epel (2012). "Telomeres and adversity: Too toxic to ignore." *Nature* 490: 169-171.

Blaxter, M. (2003). "Two worms are better than one." *Nature* 426: 395-396.

Bray, D. (2003). "Molecular Prodigality." *Science* 299: 1189-1190.

Brodin, P., V. Jojic, et al. (2015). "Variation in the human immune system is largely driven by non-heritable influences." *Cell* 160(1): 37-47.

Butler, Jason, M., Daniel J. Nolan, et al. (2010). "Endothelial Cells Are Essential for the Self-Renewal and Repopulation of Notch-Dependent Hematopoietic Stem Cells." *Cell Stem Cell* 6: 251-264.

Carlson, L. E., T. L. Beattie, et al. (2014). "Mindfulness-based cancer recovery and supportive-expressive therapy maintain telomere length relative to controls in distressed breast cancer survivors." *Cancer* 121(3): 476-484.

Celniker, S. E., D. A. Wheeler, et al. (2002). "Finishing a whole-genome shotgun: Release 3 of the Drosophila melanogaster euchromatic genome sequence." *Genome Biology* 3(12): 0079.1-0079.14.

Chakravarti, A. and P. Little (2003). "Nature, nurture and human disease." *Nature* 421: 412-414.

Cloud, John (2010). "Why Your DNA Isn't Your Destiny." *TIME*.

Darwin, F., Ed. (1888). *Charles Darwin: Life and Letters*. London, Murray.

Dennis, C. (2003). "Altered states." *Nature* 421: 686-688.

Ecker, Joseph (2012). "Genomics: ENCODE explained: Serving up a genome feast." *Nature* 489: 52-53.

Ezkurdia, I., D. Juan, et al. (2014). "Multiple evidence strands suggests that there may be as few as 19,000 human protein-coating genes." *Oxford Journals, Human Molecular Genetics* 23(22): 5866-5878.

Goldman, Bruce (2015). "Environment, not genes, dictates human immune variation, study finds." *ScienceDaily* (January 15, 2015).

Goodman, L. (2003). "Making a Genesweep: It's Official!" *Bio-IT World*.

Hall, Stephen (2012). "Interview: Journey to the Genetic Interior: What was once known as junk DNA, turns out to hold hidden treasures, says computational biologist Ewan Birney." *Scientific American* (October 2012): 80-82, 84.

Hayflick, Leonard (1965). "The limited in vitro lifetime of human diploid cell strains." *Experimental Cell Research* 37(3): 614-636.

Jablonka, E. and M. Lamb (1995). *Epigenetic Inheritance and Evolution: The Lamarckian Dimension.* Oxford, Oxford University Press.

Jones, P. A. (2001). "Death and methylation." *Nature* 409: 141-144.

Kaiser, Jocelyn (2005). "Endocrine Disrupters Trigger Fertility Problems in Multiple Generations." *Science* 308: 1391.

Khurana, E., Y. Fu, et al. (2013). "Integrative Annotation of Variants from 1092 Humans: Application to Cancer Genomics." *Science* 342: 64-84.

Kling, J. (2003). "Put the Blame on Methylation." *The Scientist* 27-28.

Kolata, Gina (2012). "Bits of Mystery DNA, Far From 'Junk,' Play Crucial Role." *New York Times*, September 5, 2012.

Lederberg, J. (1994). "Honoring Avery, MacLeod, and McCarty: The Team That Transformed Genetics." *The Scientist* 8: 11.

Lipton, B. H., K. G. Bensch, et al. (1991). "Microvessel Endothelial Cell Trans-differentiation: Phenotypic Characterization." *Differentiation* 46: 117-133.

Madhusoodanan, Jyoti (2014). "Human Gene Set Shrinks Again: Proteomic data suggest the human genome may encode fewer than 20,000 genes." *The Scientist* 9.

Nijhout, H. F. (1990). "Metaphors and the Role of Genes in Development." *Bioessays* 12(9): 441-446.

Ornish, Dean, M. J. Magbanua, et al. (2008). "Changes in prostate gene expression in men undergoing an intensive nutrition and lifestyle intervention." *Proceedings of the National Academy of Sciences* 105: 8369-8374.

Pearson, H. (2003). "Geneticists play the numbers game in vain." *Nature* 423: 576.

Pennisi, E. (2003a). "A Low Number Wins the GeneSweep Pool." *Science* 300: 1484.

Pennisi, E. (2003b). "Gene Counters Struggle to Get the Right Answer." *Science* 301: 1040-1041.

Powell, Kendall (2005). "Stem-cell niches: It's the ecology, stupid!" *Nature* 435: 268-270.

Pray, L. A. (2004). "Epigenetics: Genome, Meet Your Environment." *The Scientist* 14-20.

Preidt, Robert (2015). "Environment Trumps Genes at Shaping Immune System: Study." *U.S. News & World Report.*

Reik, W. and J. Walter (2001). "Genomic Imprinting: Parental Influence on the Genome." *Nature Reviews Genetics* 2: 21+.

Schmucker, D., J. C. Clemens, et al. (2000). "Drosophila Dscam Is an Axon Guidance Receptor Exhibiting Extraordinary Molecular Diversity." *Cell* 101: 671-684.

Seppa, N. (2000). "Silencing the BRCA1 gene spells trouble." *Science News* 157: 247.

Silverman, P. H. (2004). "Rethinking Genetic Determinism: With only 30,000 genes, what is it that makes humans human?" *The Scientist* 32-33.

Stetka, Bret (2014). "Changing our DNA through Mind Control?" *Scientific American* (December 16, 2014).

Strohman, Richard C. (2003). "Genetic determinism as a failing paradigm in biology and medicine: implications for health and wellness." *Journal of Social Work Education* 39: 169-192.

Surani, M. A. (2001). "Reprogramming of genome function through epigenetic inheritance." *Nature* 414: 122+.

Tsong, T. Y. (1989). "Deciphering the language of cells." *Trends in Biochemical Sciences* 14: 89-92.

Waterland, R. A. and R. L. Jirtle (2003). "Transposable Elements: Targets for Early Nutritional Effects on Epigenetic Gene Regulation." *Molecular and Cell Biology* 23(15): 5293-5300.

Watson, J. D., F. H. C. Crick (1953). "Molecular Structure of Nucleic Acids: A Structure for Deoxyribose Nucleic Acid." *Nature* 171: 737-738.

Watters, Ethan (2006). "DNA Is Not Destiny" *Discover*.

Willett, W. C. (2002). "Balancing Life-Style and Genomics Research for Disease Prevention." *Science* 296: 695-698.

U.S. Department of Health and Human Services (2005). *Inside the Cell: An Owner's Guide to the Cell*. Nucleus, The Cell's Brain, page 7. NIH Publication No. 05-1051 Revised September 2005.

Chapter Three

Cornell, B. A., V. L. B. Braach-Maksvytis, et al. (1997). "A biosensor that uses ion-channel switches." *Nature* 387: 580-583.

Holthuis, J. and A. Menon (2014). "Lipid landscapes and pipelines in membrane homeostasis." *Nature* 510: 48-57.

Korade, Z. and A. Kenworthy (2008). "Lipid rafts, cholesterol, and the brain." *Neuropharmacology* 55(8): 1265-1273.

Lorgeril, M., P. Salen, et al. (2010). "Cholesterol Lowering, Cardiovascular Diseases, and the Rosuvastatin-JUPITER Controversy: A Critical Reappraisal." *JAMA Network, JAMA Internal Medicine* 170(12) : 1032-1036.

Pai, V. P., S. Aw, et al. (2011). "Transmembrane voltage potential controls embryonic eye patterning in Xenopus laevis." *Development* 139(2): 313-323.

Ridker, P. M., E. Danielson, et al. (2008). "Rosuvastatin to Prevent Vascular Events in Men and Women with Elevated C-Reactive Protein." *New England Journal of Medicine* 359(21): 2195-2207.

Sultan, S. and N. Hynes (2013). "The Ugly Side of Statins. Systemic Appraisal of the Contemporary Un-Known Unknowns." *Open Journal of Endocrine and Metabolic Diseases* 3(3): 179-185.

Tsong, T. Y. (1989). "Deciphering the language of cells." *Trends in Biochemical Sciences* 14: 89-92.

Wang, K. Y., A. Tanimoto, et al. (2011). "Histamine Deficiency Decreases Atherosclerosis and Inflammatory Response in Apolipoprotein E Knockout Mice Independently of Serum Cholesterol Level." *Arterioscler Thromb Vasc Biol.* 31(4): 800-807.

Yuhas, Daisy (2013). "It's Electric: Biologists Seek to Crack Cell's Bioelectric Code." *Scientific American* (March 27, 2013).

Chapter Four

Anderson, G. L., H. L. Judd, et al. (2003). "Effects of Estrogen Plus Progestin on Gynecologic Cancers and Associated Diagnostic Procedures: The Women's Health Initiative Randomized Trial." *Journal of the American Medical Association* 290(13): 1739-1748.

Arndt, M., T. Juffmann, and V. Vedral (2009). "Quantum physics meets biology." *HFSP Journal, Frontiers of Interdisciplinary Research in the Life Sciences* 3(6): 386-400.

Balabin, I. and J. Onuchic (2000). "Dynamically Controlled Protein Tunneling Paths in Photosynthetic Reaction Centers." *Science* 290: 114-117.

Barry, Patrick (2008). "It's the Network, Stupid." *Science News* 173.

Bath, Philip and M. W., Laura J. Gray (2005). "Association between hormone replacement therapy and subsequent stroke: a meta-analysis." *British Medical Journal* 330: 342-345.

Blackman, C. F., S. G. Benane, et al. (1993). "Evidence for direct effect of magnetic fields on neurite outgrowth." Federation of American Societies for Experimental Biology 7: 801-806.

Blank, M. (1992). Na,K-ATPase function in alternating electric fields. 75th Annual Meeting of the Federation of American Societies for Experimental Biology, April 23, Atlanta, Georgia.

Burr, H. and F. Northrop (1939). "Evidence for the Existence of an Electro-Dynamic Field in Living Organisms." *PNAS* 25(6): 284-288.

Cauley, J. A., J. Robbins, et al. (2003). "Effects of Estrogen Plus Progestin on Risk of Fracture and Bone Mineral Density: The Women's Health Initiative Randomized Trial." *Journal of the American Medical Association* 290(13): 1729-1738.

Chaban, V., T. Cho., et al. (2013). "Physically disconnected non-diffusible cell-to-cell communication between neuroblastoma SH-SY5Y and DRG primary sensory neurons." *Am J Transl Res* 5(1): 69-79.

Chapman, M. S., C. R. Ekstrom, et al. (1995). "Optics and Interferometry with Na2 Molecules." *Physical Review Letters* 74(24): 4783-4786.

Chergui, M. (2006). "Controlling Biological Functions." *Science* 313: 1246-1247.

Chu, S. (2002). "Cold atoms and quantum control." *Nature* 416: 206-210.

Enserink, Martin (2010). "Newsmaker Interview: Luc Montagnier, French Nobelist Escapes 'Intellectual Terror' to Pursue Radical Ideas in China." *Science* 330: 1732.

Fenno, L., O. Yizhar, and K. Deisseroth (2011). "The development and application of optogenetics." *Annu Rev Neurosc* 34: 389-412.

Gagliano, M. and M. Renton (2013). "Love thy neighbour: facilitation through an alternative signalling modality in plants." *BMC Ecology* 13(1): 19.

Gaidos, Susan and Nicolle Rager Fuller (2009). "Living physics: From green leaves to bird brains, biological systems may exploit quantum phenomena." *Science News* 175: 26-29.

George, Alison (2006). "Lone Voices." *New Scientist* 192(2581): 44-45.

Giot, L., J. S. Bader, et al. (2003). "A Protein Interaction Map of Drosophila melanogaster." *Science* 302: 1727+.

Goodman, R. and M. Blank (2002). "Insights Into Electromagnetic Interaction Mechanisms." *Journal of Cellular Physiology* 192: 16-22.

Guan, W. and M. Reed (2012). "Electric Field Modulation of the Membrane Potential in Solid-State Ion Channels." *Nano Letters, ACS Publications* 12(12): 6441-6447.

Hackermüller, L., S. Uttenthaler, et al. (2003). "Wave Nature of Biomolecules and Fluorofullerenes." *Physical Review Letters* 91(9): 090408-1.

Hallett, M. (2000). "Transcranial magnetic stimulation and the human brain." *Nature* 406: 147-150.

Helmuth, L. (2001). "Boosting Brain Activity From The Outside In." *Science* 292: 1284-1286.

Henry, Richard Conn (2005). The mental universe. *Nature* 436: 29.

Iversen, L., H.-L. Tu, et al. (2014). "Ras activation by SOS: Allosteric regulation by altered fluctuation dynamics." *Science* 345: 50-54.

Jansen, R., H. Yu, et al. (2003). "A Bayesian Networks Approach for Predicting Protein-Protein Interactions from Genomic Data." *Science* 302: 449-453.

Jin, M., M. Blank, et al. (2000). "ERK1/2 Phosphorylation, Induced by Electromagnetic Fields, Diminishes During Neoplastic Transformation." *Journal of Cell Biology* 78: 371-379.

Josephson, Brian (2004). "Pathological Disbelief." Lecture given at the Nobel Laureate Meeting, Lindau, June 30th, 2004.

Kesari, K. K., M. H. Siddiqui, et al. (2013). "Cell Phone Radiation Exposure on Brain and Associated Biological Systems." *Indian Journal of Experimental Biology* 51: 187-200.

Kübler-Ross, Elizabeth (1997) *On Death and Dying*, New York, Scribner.

Li, S., C. M. Armstrong, et al. (2004). "A Map of the Interactome Network of the Metazoan C. elegans." *Science* 303: 540+.

Liboff, A. R. (2004). "Toward an Electromagnetic Paradigm for Biology and Medicine." *Journal of Alternative and Complementary Medicine* 10(1): 41-47.

Lipton, B. H., K. G. Bensch, et al. (1991). "Microvessel Endothelial Cell Transdifferentiation: Phenotypic Characterization." *Differentiation* 46: 117-133.

McClare, C. W. F. (1974). "Resonance in Bioenergetics." *Annals of the New York Academy of Sciences* 227: 74-97.

Nahin, R. L., P. M. Barnes, et al. (2009). "Costs of Complementary and Alternative Medicine (CAM) and Frequency of Visits to CAM Practitioners: United States, 2007." *National health statistics reports; no. 18*. Hyattsville, MD, National Center for Health Statistics.

Null, G., Ph.D., C. Dean, M.D. N.D., et al. (2003). *Death By Medicine*. New York, Nutrition Institute of America.

Oschman, J. L. (2000). Chapter 9: Vibrational Medicine. *Energy Medicine: The Scientific Basis*. Edinburgh, Harcourt Publishers: 121-137.

Pagels, H. R. (1982). *The Cosmic Code: Quantum Physics As the Language of Nature*. New York, Simon and Schuster.

Pool, R. (1995). "Catching the Atom Wave." *Science* 268: 1129-1130.

Pophristic, V. and L. Goodman (2001). "Hyperconjugation not steric repulsion leads to the staggered structure of ethane." *Nature* 411: 565-568.

Prokhorenko, V. I., A. M. Nagy, et al. (2006). "Coherent Control of Retinal Isomerization in Bacteriorhodopsin." *Science* 313: 1257-1261.

Richards, G. H., K. E. Wilk, et al. (2012). "Coherent Vibronic Coupling in Light-Harvesting Complexes from Photosynthetic Marine Algae." *The Journal of Physical Chemistry Letters* 3(2): 272-277.

Rosen, A. D. (1992). "Magnetic field influence on acetylcholine release at the neuromuscular junction." *American Journal of Physiology-Cell Physiology* 262: C1418-C1422.

Rumbles, G. (2001). "A laser that turns down the heat." *Nature* 409: 572-573.

Sanders, Laura (2014). "Brain Hack: Consumers take their neurons into their own hands." *Science News* 186(10) (November 15, 2014): 22-25.

Sarovar, M., A. Ishizaki, et al. (2010.) "Quantum entanglement in photosynthetic light-harvesting complexes." *Nature Physics* 6: 462-467.

Schulten, Klaus (2000). "Electron Transfer: Exploiting Thermal Motion." *Science* 290: 61-62.

Shumaker, S. A., C. Legault, et al. (2003). "Estrogen Plus Progestin and the Incidence of Dementia and Mild Cognitive Impairment in Postmenopausal Women: The Women's Health Initiative Memory Study: A Randomized Controlled Trial." *Journal of the American Medical Association* 289(20): 2651-2662.

Sivitz, L. (2000). "Cells proliferate in magnetic fields." *Science News* 158: 195.

Starfield, B. (2000). "Is US Health Really the Best in the World?" *Journal of the American Medical Association* 284(4): 483-485.

Szent-Györgyi, A. (1960). *Introduction to a Submolecular Biology.* New York, Academic Press.

Tsong, T. Y. (1989). "Deciphering the language of cells." *Trends in Biochemical Sciences* 14: 89-92.

Valone, Thomas (2000). *Bioelectromagnetic Healing: A Rationale for Its Use.* Maryland, Integrity Research Institute.

Wassertheil-Smoller, S., S. L. Hendrix, et al. (2003). "Effect of Estrogen Plus Progestin on Stroke in Postmenopausal Women: The Women's Health Initiative: A Randomized Trial." *Journal of the American Medical Association* 289(20): 2673-2684.

Weinhold, F. (2001). "A new twist on molecular shape." *Nature* 411: 539-541.

Yen-Patton, G. P. A., W. F. Patton, et al. (1988). "Endothelial Cell Response to Pulsed Electromagnetic Fields: Stimulation of Growth Rate and Angiogenesis In Vitro." *Journal of Cellular Physiology* 134: 37-46.

Zukav, G. (1979). *The Dancing Wu Li Masters: An Overview of the New Physics.* New York, Bantam.

Chapter Five

Achor, Shawn (2010). *The Happiness Advantage: The Seven Principles of Positive Psychology That Fuel Success and Performance at Work.* New York, Crown Business.

Benson, Herbert and Marg Stark (1997). *Timeless Healing: The Power and Biology of Belief.* New York, Scribner.

Brown, W. A. (1998). "The Placebo Effect: Should doctors be prescribing sugar pills?" *Scientific American* 278(1): 90-95.

Burton, Claire L., S. Chhabra, et al. (2002). "The Growth Response of Escherichia coli to Neurotransmitters and Related Catecholamine Drugs Requires a Functional Enterobactin Biosynthesis and Uptake System." *Infection and Immunology* 70: 5913-5923.

Cole, S. W., M. E. Kemeny, et al. (1996). "Accelerated course of human immunodeficiency virus infection in gay men who conceal their homosexual identity." *Psychosomatic Medicine* 58(3): 219-231.

Cole, Steve (2009). "Social Regulation of Human Gene Expression." *Sage Current Directions in Psychological Science* 18(3): 132-137.

Crum, A. and E. Langer (2007). "Mind-set matters: Exercise and the placebo effect." *Psychological Science* 18(2): 165-171.

Dobbs, David (2013). "Feature Story: The Social Life of Genes." *Pacific Standard* magazine (September/October 2013).

DiRita, V. J. (2000). "Genomics Happens." *Science* 289: 1488-1489.

Discovery (2003). *Placebo: Mind Over Medicine? Medical Mysteries.* Silver Spring, MD, Discovery Health Channel.

Erdmann, J. (2008). "Imagination Medicine." *Science News* 174: 26-30.

Greenberg, G. (2003). "Is It Prozac? Or Placebo?" *Mother Jones:* 76-81.

Grierson, Bruce (2014). "What If Age Is Nothing but a Mindset?" *New York Times,* October 22, 2014, Magazine, The Health Issue.

Horgan, J. (1999). Chapter 4: Prozac and Other Placebos. *The Undiscovered Mind: How the Human Brain Defies Replication, Medication, and Explanation.* New York, The Free Press: 102-136.

Kaliman, P., M. J. Alvarez-Lopez, et al. (2014). "Rapid changes in histone deacetylases and inflammatory gene expression in expert meditators." *Psychoneuroendocrinology* 40: 96-107.

Kaufman, J., B. Z. Yang, et al. (2004). "Social supports and serotonin transporter gene moderate depression in maltreated children." *PNAS* 101(49): 17316-17321.

Kawashima, Koichiro, H. Misawa, et al. (2007)."Ubiquitous expression of acetylcholine and its biological functions in life forms without nervous systems." *Life Sciences* 80: 2206-2209.

Kirsch, I., T. J. Moore, et al. (2002). "The Emperor's New Drugs: An Analysis of Antidepressant Medication Data Submitted to the U.S. Food and Drug Administration." *Prevention & Treatment* (American Psychological Association) 5: Article 23.

Langer, Ellen (2009). *Counter Clockwise: Mindful Health and the Power of Possibility.* New York, Ballantine Books.

Lee, R. S., K. L. Tamashiro, et al. (2010). "Chronic Corticosterone Exposure Increases Expression and Decreases Deoxyribonucleic Acid Methylation of Fkbp5 in Mice." *Endocrinology* 151(9): 4332-4343.

Leuchter, A. F., I. A. Cook, et al. (2002). "Changes in Brain Function of Depressed Subjects During Treatment With Placebo." *American Journal of Psychiatry* 159(1): 122-129.

Levy, N. R., C. Pilver, et al. (2014). "Sublimal Strengthening: Improving Older Individuals' Physical Function over Time with an Implicit-Age-Stereotype Intervention." *Sage Journals, Psychological Science* 25(12): 2127-2135.

Lipton, B. H., K. G. Bensch, et al. (1992). "Histamine-Modulated Transdifferentiation of Dermal Microvascular Endothelial Cells." *Experimental Cell Research* 199: 279-291.

Mason, A. A. (1952). "A Case of Congenital Ichthyosiform Erythrodermia of Brocq Treated by Hypnosis." *British Medical Journal* 30: 442-443.

Miller, G., N. Rohleder, and S. Cole (2009). "Chronic interpersonal stress predicts activation of pro- and anti-inflammatory signaling pathways six months later." *Psychosomatic Medicine* 71(1): 57-62.

Moseley, J. B., K. O'Malley, et al. (2002). "A Controlled Trial of Arthroscopic Surgery for Osteoarthritis of the Knee." *New England Journal of Medicine* 347(2): 81-88.

Niemi, Maj-Britt (2009). "Placebo Effect: A Cure in the Mind." *Scientific American Mind* 20: 42-49.

Naokuni, Takeda and S. Kanji (1993). "Metabolism of biogenic monoamines in the ciliated protozoan, Tetrahymena pyriformis." *Comparative Biochemistry and Physiology* Part C 106: 63-70.

Paul, Gordon (1963). "The Production of Blisters by Hypnotic Suggestion: Another Look." *Psychosomatic Medicine* 25(3): 233-244.

Pert, Candace (1997). *Molecules of Emotion: The Science Behind Mind-Body Medicine*, New York, Scribner.

Price, D. D., D. G. Finniss, et al. (2008). "A Comprehensive Review of the Placebo Effect: Recent Advances and Current Thought." *Annual Review of Psychology* 59: 565-590.

Ryle, G. (1949). *The Concept of Mind*. Chicago, University of Chicago Press.

Szegedy-Maszak, Marianne (2005). "Mysteries of the Mind: Your unconscious is making your everyday decisions." *U.S. News & World Report*.

Chapter Six

Ackerman, Diane (2012). "The Brain on Love." *The New York Times*, March 24, 2012, The Opinion Pages.

Atkinson, William (2000). "Strategies for Workplace Stress." Risk & Insurance Online (www.riskandinsurance.com).

Arnsten, A. F. T. and P. S. Goldman-Rakic (1998). "Noise Stress Impairs Prefrontal Cortical Cognitive Function in Monkeys: Evidence for a Hyperdopaminergic Mechanism." *Archives of General Psychiatry* 55: 362-368.

Bhasin, M. K., J. A. Dusek, et al. (2013). "Relaxation Response Induces Temporal Transcriptome Changes in Energy Metabolism, Insulin Secretion and Inflammatory Pathways." *PLOS ONE* 8(5):e62817.

Chetty, S., A. R. Friedman, et al. (2014). "Stress and glucocorticoids promote oligodendrogenesis in the adult hippocampus." *Molecular Psychiatry* 19(12): 1275-1283.

Coan, J., H. Schaefer, and R. Davidson (2006). "Lending a Hand: Social Regulation of the Neural Response to Threat." *Sage Journals, Psychological Science* 17(12): 1032-1039.

Cohen, S., D. Janicki-Deverts, et al. (2012). "Chronic stress, glucocorticoid receptor resistance, inflammation, and disease risk." *PNAS* 109(16): 5995-5999.

Goldstein, L. E., A. M. Rasmusson, et al. (1996). "Role of the Amygdala in the Coordination of Behavioral, Neuroendocrine, and Prefrontal Cortical Monoamine Responses to Psychological Stress in the Rat." *Journal of Neuroscience* 16(15): 4787-4798.

Holden, C. (2003). "Future Brightening for Depression Treatments." *Science* 302: 810-813.

Kopp, M. S. and J. Réthelyi (2004). "Where psychology meets physiology: chronic stress and premature mortality—the Central-Eastern European health paradox." *Brain Research Bulletin* 62: 351-367.

Lewis, Scarlett (2014). *Nurturing Healing Love: A Mother's Journey of Hope and Forgiveness.* Carlsbad, Hay House, Inc.

Lipton, B. H., K. G. Bensch, et al. (1991). "Microvessel Endothelial Cell Transdifferentiation: Phenotypic Characterization." *Differentiation* 46: 117-133.

McEwen, B. S. and T. Seeman (1999). "Protective and Damaging Effects of Mediators of Stress: Elaborating and Testing the Concepts of Allostasis and Allostatic Load." *Annals of the New York Academy of Sciences* 896: 30-47.

McEwen, B. and with Elizabeth N. Lasley (2002). *The End of Stress As We Know It.* Washington, National Academies Press.

Norman, L., N. Lawrence, et al. (2014). "Attachment-security priming attenuates amygdala activation to social and linguistic threat." *Social Cogn & Affect Neurosci* 127.

Segerstrom, S. C. and G. E. Miller (2004). "Psychological Stress and the Human Immune System: A Meta-Analytic Study of 30 Years of Inquiry." *Psychological Bulletin* 130(4): 601-630.

Takamatsu, H., A. Noda, et al. (2003). "A PET study following treatment with a pharmacological stressor, FG7142, in conscious rhesus monkeys." *Brain Research* 980: 275-280.

Van Engen, N. K., M. L. Stock, et al. (2014). "Impact of oral meloxicam on circulating physiological biomarkers of stress and inflammation in beef steers after long-distance transportation." *Journal of Animal Science* 92(2): 498-510.

Weich, S., H. L. Pearce, et al. (2014). "Effect of anxiolytic and hypnotic drug prescriptionson mortality hazards: retrospective cohort study." *BMJ* 348:g1996.

Chapter Seven

Almgren, M., T. Schlinzig, et al. (2014). "Cesarean delivery and hemato-poietic stem cell epigenetics in the newborn infant: implications for future health?" *American Journal of Obstetrics & Gynecology* 211(5): 502.e1-502.e8.

Arnsten, A. F. T. (2000). "The Biology of Being Frazzled." *Science* 280: 1711-1712.

Bateson, P., D. Barker, et al. (2004) "Developmental plasticity and human health." *Nature* 430: 419-421

Bhattacharjee, Yudhijit (2015). "Baby Brains." *National Geographic Magazine* (January 2015).

Blaser, Martin J. M.D. (2014). *Missing Microbes: How the Overuse of Antibiotics Is Fueling Our Modern Plagues*. New York, Henry Holt.

Burkitt, Laurie (2014). "In China, Expectant Dads Line Up to Experience Labor Pains." *The Wall Street Journal* (December 18, 2014).

Centers for Disease Control and Prevention (2013). *Antibiotic Resistance Threats in the United States, 2013*. Atlanta, CDC.

Chamberlain, D. (1998). *The Mind of Your Newborn Baby*. Berkeley, CA, North Atlantic Books.

Christensen, D. (2000). "Weight Matters, Even in the Womb: Status at birth can foreshadow illnesses decades later." *Science News* 158: 382-383.

Daniels, S., F. Greer, and the Committee on Nutrition (2008). "Lipid Screening and Cardiovascular Health in Childhood." *Pediatrics* 122(1): 198-208.

Devlin, B., M. Daniels, et al. (1997). "The heritability of IQ." *Nature* 388: 468-471.

Dodic, M., V. Hantzis, et al. (2002). "Programming effects of short prenatal exposure to cortisol." *Federation of American Societies for Experimental Biology* 16: 1017-1026.

Gibbons, Ann (1998). "Solving the Brain's Energy Crisis." *Science* 280: 1345-1347.

Gluckman, P. D. and M. A. Hanson (2004). "Living with the Past: Evolution, Development, and Patterns of Disease." *Science* 305: 1733-1736.

Gunnar, M. and K. Quevedo (2008). "Early care experiences and HPA axis regulation in children: a mechanism for later trauma vulnerability." *Progress in Brain Research* 167: 137-149.

Hart, B. and T. Risley (2003). "The Early Catastrophe: The 30 Million Word Gap by Age 3." *American Educator* Spring 2003: 4-9.

Holden, C. (1996). "Child Development: Small Refugees Suffer the Effects of Early Neglect." *Science* 274(5290): 1076-1077.

Jakobsson, H. E., T. R. Abrahamsson, et al. (2014). "Decreased gut microbiota diversity, delayed Bacteroidetes colonization and reduced Th1 responses in infants delivered by Caesarean section." *Gut* 63(4): 559-566.

Kuhl, Patricia (2011). "Early Language Learning and Literacy: Neuroscience Implications for Education." *Journal Compilation 2011 International Mind, Brain, and Education Society and Blackwell Publishing, Inc.* 5(3): 128-142.

Laibow, R. (1999). *Clinical Applications: Medical applications of neurofeedback. Introduction to Quantitative EEG and Neurofeedback.* J. R. Evans and A. Abarbanel. Burlington, MA, Academic Press (Elsevier).

Laibow, R. (2002). Personal communication with B. H. Lipton. New Jersey.

Leake, Jonathan (2007). "Love of broccoli begins in the womb." *The Sunday Times* (December 2, 2007).

Lesage, J., F. Del-Favero, et al. (2004). "Prenatal stress induces intrauterine growth restriction and programmes glucose intolerance and feeding behaviour disturbances in the aged rat." *Journal of Endocrinology* 181: 291-296.

Leutwyler, K. (1998). "Don't Stress: It is now known to cause developmental problems, weight gain and neurodegeneration." *Scientific American* 278(1): 28-30.

Lewin, R. (1980). "Is Your Brain Really Necessary?" *Science* 210: 1232-1234.

Maguire, E. A., D. G. Gadian, et al. (2000). "Navigation-related structural change in the hippocampi of taxi drivers." *PNAS* 97(8): 4398-4403.

McEwen, B. and P. Gianaros (2010). "Central role of the brain in stress and adaptation: links to socioeconomic status, health, and disease." *Ann N Y Acad Sci* 1186: 190-222.

McGue, M. (1997). "The democracy of the genes." *Nature* 388: 417-418.

Mendizza, M. and J. C. Pearce (2001). *Magical Parent, Magical Child.* Nevada City, CA, Touch the Future.

Nathanielsz, P. W. (1999). *Life In the Womb: The Origin of Health and Disease.* Ithaca, NY, Promethean Press.

Norretranders, T. (1998). *The User Illusion: Cutting Consciousness Down to Size.* New York, Penguin Books.

Prescott, J. W. (1990). *Affectional Bonding for the Prevention of Violent Behaviors: Neurobiological, Psychological and Religious/Spiritual Determinants.* Violent Behaviour, Volume I: Assessment & Intervention. L. J. Hertzberg, G. F. Ostrum and J. R. Field. Great Neck, NY, PMA Publishing Corp. One: 95-125.

Prescott, J. W. (1996). "The Origins of Human Love and Violence." *Journal of Prenatal & Perinatal Psychology & Health* 10(3): 143-188.

Reik, W. and J. Walter (2001). "Genomic Imprinting: Parental Influence on the Genome." *Nature Reviews Genetics* 2: 21+.

Sandman, C. A., P. D. Wadhwa, et al. (1994). "Psychobiological Influences of Stress and HPA Regulation on the Human Fetus and Infant Birth Outcomes." *Annals of the New York Academy of Sciences* 739 (Models of Neuropeptide Action): 198-210.

Sapolsky, R. M. (1997). "The Importance of a Well-Groomed Child." *Science* 277: 1620-1621.

Shonkoff, Jack P., W. Thomas Boyce, and Bruce S. McEwen (2009). "Neuroscience, Molecular Biology, and the Childhood Roots of Health Disparities." *Journal of the American Medical Association* 301: 2252-2259.

Schultz, E. A. and R. H. Lavenda (1987). *Cultural Anthropology: A Perspective on the Human Condition*. St. Paul, MN, West Publishing. *Science* (2001). "Random Samples." *Science* 292(5515): 205+.

Science, Editorial Staff (2001). Random Samples: "Like Mother, Like Son." *Science* 292: 205.

Siegel, D. J. (1999). *The Developing Mind: How Relationships and the Brain Interact to Shape Who We Are*. New York, Guilford.

Surani, M. A. (2001). "Reprogramming of genome function through epigenetic inheritance." *Nature* 414: 122+.

Suskind, D., K. R. Leffel, et al. (2013). "An Exploratory Study of 'Quantitative Linguistic Feedback.'" *Sage Journals, Communication Disorders Quarterly* 34(4): 199-209.

Vaiserman, Alexander (2014). "Early-life Exposure to Endocrine Disrupting Chemicals and Later-life Health Outcomes: An Epigenetic Bridge?" *Aging and Disease* 5(6): 419-429.

Victora, C. G., P.G. Smith, et al. (1989). "Infant feeding and deaths due to diarrhea. A case-control study." *Oxford Journals, Am J Epidemiol* 129(5): 1032-1041.

Verny, T. and with John Kelly (1981). *The Secret Life of the Unborn Child*. New York, Bantam Doubleday Dell.

Verny, T. R. and Pamela Weintraub (2002). New York, Simon & Schuster.

Wong, Sam (2014). "Imperial College London: Caesarean babies more likely to become overweight as adults, analysis finds." *ScienceDaily* (February 26, 2014).

Zhong, W., H. Maradit-Kremers, et al. (2013). "Age and Sex Patterns of Drug Prescribing in a Defined American Population." *Mayo Clinic Proc.* 88(7): 697-707.

Epilogue

Alexander, Eben M.D. (2012). *Proof of Heaven: A Neurosurgeon's Journey into the Afterlife*. New York, Simon & Schuster.

deWaal, F. B. M. (2004). "Peace Lessons from an Unlikely Source." *Public Library of Science—Biology* 2(4): 0434-0436.

Gould, Stephen Jay (2002). *Rocks of Ages: Science and Religion in the Fullness of Life*. New York, Ballantine Books.

Hoffman, B. and H. Dukas (1972). *Albert Einstein: Creator and Rebel*. London, Hart-Davis, MacGibbon.

Koenig, Harold (2012). "Religion, Spirituality, and Health: The Research and Clinical Implications." *ISRN Psychiatry* 2012, Article ID 278730.

Lanza, Robert M.D. "Does Death Exist? New Theory Says 'No.'" *Huffington post.com*, Nov. 17, 2011.

Leininger, A., B. Leininger, and K. Gross (2009). *Soul Survivor: The Reincarnation of a World War II Fighter Pilot*. New York, Grand Central Publishing.

Mayr, E. (1976). *Evolution and the Diversity of Life: Selected Essays*. Cambridge, Harvard University Press.

Moorjani, Anita (2012). *Dying to Be Me: My Journey from Cancer, to Near Death, to True Healing*. Carlsbad, Hay House, Inc.

Pearsall, P. (1998). *The Heart's Code: Tapping the Wisdom and Power of Our Heart Energy*. New York, Random House.

Rendu, W., C. Beauval, et al. (2014). "Evidence supporting an intentional Neandertal burial at La-Chapelle-aux-Saints." *PNAS* 111(1): 81-86.

Sapolsky, R. M. and L. J. Share (2004). "A Pacific Culture among Wild Baboons: Its Emergence and Transmission." *Public Library of Science—Biology* 2(4): 0534-0541.

Sylvia, C. and W. Novak (1997). *A Change of Heart: A Memoir*. Boston, Little, Brown and Company.

Addendum

Fannin, J. L. and R. M. Williams (2012). "Neuroscience Reveals the Whole-Brain State and Its Applications for International Business and Sustainable Success." *IJMB* 3(1): 73-95

Index

(Page numbers in italics indicate an illustration)

259

About the Author

Bruce Lipton, Ph.D., is an internationally recognized authority in bridging science and spirit. He has been a guest speaker on hundreds of TV and radio shows as well as keynote presenter for national conferences.

Dr. Lipton began his scientific career as a cell biologist. He received his Ph.D. from the University of Virginia at Charlottesville before joining the Department of Anatomy at the University of Wisconsin's School of Medicine in 1973. Dr. Lipton's research on muscular dystrophy, studies employing cloned human stem cells, focused upon the molecular mechanisms controlling cell behavior. An experimental tissue transplantation technique developed by Dr. Lipton and colleague Dr. Ed Schultz and published in the journal *Science* was subsequently employed as a novel form of human genetic engineering.

In 1982, Dr. Lipton began examining the principles of quantum physics and how they might be integrated into his understanding of the cell's information processing systems. He produced breakthrough studies on the cell membrane, which revealed that this outer layer of the cell was an organic homologue of a computer chip, the cell's equivalent of a brain. His research at Stanford University's School of Medicine, between 1987 and 1992, revealed that the environment, operating though the membrane, controlled the behavior and physiology of the cell, turning genes on and off. His discoveries, which ran counter to the established scientific view that life is controlled by the genes, presaged one of today's most important fields of study, the science of epigenetics. Two major scientific publications derived from these studies defined the molecular pathways connecting the mind and body. Many subsequent papers by other researchers have since validated his concepts and ideas.

Dr. Lipton has taken his award-winning medical school lectures to the public and is currently a sought after keynote speaker and workshop presenter. He lectures to conventional and complementary medical professionals and lay audiences about leading-edge science and how it dovetails with mind-body medicine and spiritual principles. He has been heartened by anecdotal reports from hundreds of former audience members who have improved their spiritual, physical, and mental well-being by applying the principles he discusses in his lectures. He is regarded as one of the leading voices of the New Biology.

We hope you enjoyed this Hay House book. If you'd like to receive our online catalog featuring additional information on Hay House books and products, or if you'd like to find out more about the Hay Foundation, please contact:

Hay House, Inc., P.O. Box 5100, Carlsbad, CA 92018-5100
(760) 431-7695 or (800) 654-5126
(760) 431-6948 (fax) or (800) 650-5115 (fax)
www.hayhouse.com® • www.hayfoundation.org

———

Published in Australia by: Hay House Australia Pty. Ltd.,
18/36 Ralph St., Alexandria NSW 2015
Phone: 612-9669-4299 • *Fax:* 612-9669-4144
www.hayhouse.com.au

Published in the United Kingdom by: Hay House UK, Ltd.,
The Sixth Floor, Watson House, 54 Baker Street, London W1U 7BU
Phone: +44 (0)20 3927 7290 • *Fax:* +44 (0)20 3927 7291
www.hayhouse.co.uk

Published in India by: Hay House Publishers India,
Muskaan Complex, Plot No. 3, B-2, Vasant Kunj, New Delhi 110 070
Phone: 91-11-4176-1620 • *Fax:* 91-11-4176-1630
www.hayhouse.co.in

———

Access New Knowledge.
Anytime. Anywhere.

Learn and evolve at your own pace
with the world's leading experts.

www.hayhouseU.com

Listen. Learn. Transform.

Embrace vibrant, lasting health with unlimited Hay House audios!

Unlock endless wisdom, fresh perspectives, and life-changing tools from world-renowned authors and teachers—helping you live your happiest, healthiest life. With the *Hay House Unlimited Audio* app, you can learn and grow in a way that fits your lifestyle . . . and your daily schedule.

With your membership, you can:

- Develop a healthier mind, body, and spirit through natural remedies, healthy foods, and powerful healing practices.

- Explore thousands of audiobooks, meditations, immersive learning programs, podcasts, and more.

- Access exclusive audios you won't find anywhere else.

- Experience completely unlimited listening. No credits. No limits. No kidding.

Try for FREE!